Contents

I0181579

The Power of Reflection

Nate Needham

Copyright

The Power of Reflection

Ancient Wisdom Meets Modern
Science to Relieve Stress,
Strengthen Relationships, and
Inspire the Future

Nate Needham

Final Words

About the Author

Nate Needham is an entrepreneur, author, and the founder of Fiboprana Publishing, a creative company dedicated to exploring the connection between science, consciousness, and the natural intelligence of life. Through his books, Nate shares insights that bridge ancient wisdom and modern understanding, helping readers live with greater clarity, energy, and purpose.

For nearly 15 years, Nate has followed an unconventional path as both a business owner and lifelong student of human potential. His personal journey, marked by deep reflection, challenge, and discovery, led him to see how perception shapes experience and how awareness can transform the way we live. This realization inspired the creation of Fiboprana Publishing, whose mission is to make timeless truths simple, practical, and accessible for the modern world.

Each Fiboprana book reflects a different facet of that vision: awakening the mind, restoring balance to the body, and reconnecting people with the natural harmony that already exists within and around them. Nate's work is guided by one central idea: that by understanding ourselves more deeply, we can unlock the limitless intelligence of life itself.

Preface

This is not a book of quick fixes or surface-level advice. It asks you to look at something most of us overlook: the mirror of your own mind.

At first, this idea may feel strange. We live in a world that tells us to solve problems by changing the outside: our jobs, our relationships, our achievements, our possessions. We're taught to chase validation, improvement, or success, as if fixing the reflection will finally make us happy. Rarely do we stop to consider the mirror itself: the way we see, interpret, and reflect on everything around us.

But that mirror quietly shapes everything. When it's clouded by fear, self-doubt, or old stories, even the best opportunities can look like threats. When it's polished and clear, challenges become possibilities, relationships deepen, and life feels lighter. The outside world may look the same, but the way you experience it changes completely.

The ideas in these pages draw from two great rivers of knowledge: ancient wisdom traditions that liken the mind to water, glass, or shadow; and modern science, which demonstrates how our brains filter reality through beliefs, biases, and emotions. The language may differ, but the message converges: clarity within creates clarity without.

Some of this may feel unfamiliar at first. The notion that you are not your thoughts, that your worth is not up for debate, that happiness is already here, or that change can be easy, these may sound too simple, too radical, or too different from what you've been told. But the benefits of exploring them are profound.

Less stress and anxiety. By learning to see thoughts and emotions for what they are (passing reflections, not fixed truths), you stop fighting shadows.

Stronger relationships. When judgment and fear no longer dominate, you reflect compassion, empathy, and connection instead of defensiveness.

Resilience in the face of challenges. A clear mirror doesn't erase difficulty, but it lets you respond wisely rather than react blindly.

A sense of enough-ness. You no longer need to prove or earn your worth; you begin to rest in it.

A brighter vision of the future. Reflection expands perspective, revealing possibilities where you once saw limits.

The mirror of the mind cannot be polished once and for all. It's a lifelong practice.

But even small acts of reflection create ripples of change. And as individual mirrors clear, entire families, communities, and societies begin to shine brighter.

That is the invitation of this book: to pause, reflect, and rediscover the power that has always been within you, the power of reflection.

Introduction

For most of my life, I've walked a path different from the one Western society typically expects. I've been an entrepreneur for nearly 15 years, and in the spaces between business, I've pursued the extreme sports I love with passion and freedom. From the outside, this might have looked unconventional, even reckless to some. And over the years, I've heard many judgments and misinterpretations about the way I've chosen to live.

At first, these misjudgments stung. I felt offended, angry, and misunderstood. I tried explaining myself, reasoning with people, showing them the logic and heart behind my choices. But no matter what I said, their opinions rarely changed. The more I pushed back, the more it seemed like the world was against me. My anger and frustration grew until I began to see everything through that distorted lens as if every red light, every rejection, every doubt cast by others was proof that life was unfair.

But then something shifted. Through much reflection, I began to notice how heavy this anger had become. It wasn't just other people's words weighing on me; it was my perception of them. The more I focused on being misunderstood, the more distorted my perception of the world became. And in that realization came a choice: to let go.

I decided to forgive both those who had judged me and the shadows of judgment I had created in my own mind. I began to see that most people carry their own distortions, their own pains, and filters. Their opinions of me didn't define me. In fact, they never really could. I no longer needed validation, approval, or defense. I realized that the very "self" I was trying to protect was primarily shaped by conditioning and stories handed to me by society. And if that self could be let go, if the mirror could be cleared, then I would be free.

That realization is what inspired the creation of this book.

For thousands of years, cultures around the world have pointed to the same truth: we don't experience life directly; we experience it as a reflection of our inner state. Taoist sages compared the mind to still water: when calm, it reflects clearly; when stirred, it distorts. Buddhist teachers spoke of the mind's mirror being clouded by craving and ignorance. In ancient Greece, Plato told the story of shadows mistaken for reality until someone turned toward the light.

Modern science echoes this wisdom. Neuroscience reveals that our brains don't capture reality like a camera; instead, they filter it through memory, belief, and emotion. Psychology tells us that stress makes us see threats, while optimism reveals opportunities. Two people can face the exact same situation and live in entirely different realities, all because of the mirror they're looking through.

This book is about that mirror. It's about how easily it becomes fogged by fear, anger, self-doubt, and the relentless pursuit of external validation. It's about the practices, ancient and modern, that help us polish it clearly. And it's about the freedom that comes when we no longer live trapped inside distorted reflections.

Living with a polished mirror doesn't mean life becomes perfect. Challenges still arise, people still misunderstand, setbacks still come. But clarity allows us to see them as they are, not as fear or ego paints them. With a clear mirror, we respond with wisdom, compassion, and freedom.

That is the path I've walked into, and it's the invitation I offer you in these pages. Together, we'll uncover the misconceptions that cloud the mirror, learn practices to polish it clear, and discover what life looks like when we stop needing the world's validation and begin living from clarity within.

Because when one mirror clears, the whole world shines brighter.

Chapter 1 – Buddhism: The Mind as a Mirror

The Angry Monk and the Mirror

In a quiet monastery nestled in the hills, the monks' lives followed a simple rhythm: meditation at dawn, work in the fields, study, and prayer. It was a place meant to cultivate peace, but inside one young monk's heart, a storm was brewing. He had been at the monastery for several years, yet he constantly compared himself to others. When he saw another monk praised for his discipline, envy filled him. When he struggled to sit through long hours of meditation, frustration gnawed at him. At night, he replayed conversations in his head, craving approval and recognition, and resenting those who seemed further along the path.

One morning, unable to contain his feelings, he burst into the meditation hall where his teacher sat. His voice, usually soft, rose in anger. "I cannot do this!" he shouted. "I try, and I fail. I want peace, but my mind is full of restlessness. Others seem calm, but I am drowning in thoughts and emotions. What is the point of this practice if I cannot even control myself?"

The other monks froze in silence, waiting for the master to rebuke him. But the teacher did not frown, nor did he raise his voice. Instead, he reached beside him and lifted a small mirror, polished to a bright shine. Holding it before the young monk, he asked gently, "What do you see?"

"My reflection," the monk replied, confused.

The teacher nodded. "And if mud were smeared on this mirror, would you see clearly?"

"No," said the monk. "The image would be distorted."

The teacher lowered the mirror. "So it is with the mind. Anger, craving, and ignorance are like mud on the surface.

They do not destroy the mirror, but they cover it. When you shouted just now, you were not wrong for feeling anger, but you mistook the mud for the mirror itself. You believed the distortion was who you are." The monk's eyes widened. "Then what should I do?"

"Polish the mirror," the teacher said simply. "Not by forcing the mud away, but by patiently sitting, breathing, watching. See your thoughts and emotions for what they are: passing reflections, not the truth of who you are. Do this again and again, and you will notice: beneath the mud, the mirror has always been clear."

The young monk felt tears well up in his eyes. His anger had seemed so heavy, so permanent, but in that moment, he realized it was only a passing shadow. For the first time, he understood the heart of practice: not to become someone else, but to see himself clearly.

From that day on, whenever frustration rose, he remembered the mirror. Slowly, with breath and awareness, he learned to polish his own mind, discovering the clarity that had always been waiting beneath the clouds.

The story of the angry monk and the mirror reminds us of something profound: the mind itself is not flawed, but it often cannot see clearly because it is covered by layers of craving, ignorance, and attachment. In Buddhist thought, the mirror is always there, bright and capable of reflecting reality as it truly is. The work of practice is not to create clarity, but to uncover it, so its natural brilliance can shine through.

The Clouded Mirror

Craving is the first smear on the surface of the mind. In Buddhism, craving (tanha) is described as the thirst that drives much of human behavior: the constant desire for more pleasure, more recognition, and more certainty. When we crave, our perception becomes warped. We no longer see things as they are, but as how we wish them to be. The desire for permanence in a world of change sets us up for disappointment, just as the young monk's craving for recognition blinded him to his own progress.

Ignorance (avidyā) is considered the root of suffering. It is not just a lack of information, but a deeper misunderstanding of reality. We ignore the truth of impermanence, imagining that the people, possessions, or even identities we cling to will last forever. Ignorance also leads us to misperceive the self, mistaking passing thoughts and feelings for who we really are. This blindness keeps us locked in cycles of suffering, repeating the same mistakes because we cannot see clearly.

Emotional attachment adds another layer of fog. When we become entangled with our emotions, whether joy, sadness, anger, or fear, we mistake them for permanent truths. We identify with them, saying, "I am angry" or "I am broken," as if those states define us completely. In reality, emotions are like clouds drifting across the sky. They obscure the view, but they are not the sky itself.

Together, craving, ignorance, and attachment cloud the mirror of the mind, creating distorted reflections. To see clearly, we must learn to polish away these distortions.

Polishing the Mirror Through Practice

Buddhist traditions offer practical methods to polish the mind so it can reflect reality clearly. The first is mindfulness (sati), the simple but radical act of being fully aware of the present moment. By noticing the breath, the body, or the stream of thoughts without clinging to them, we begin to see that experiences arise and pass away naturally.

Mindfulness is like gently wiping dust from the mirror, allowing us to notice what is really there.

Insight meditation (Vipassana) takes this further by helping practitioners see impermanence directly. Sitting quietly, observing sensations, thoughts, and feelings as they come and go, we begin to understand that nothing lasts. This insight frees us from the illusion that we can hold onto pleasure forever or avoid discomfort completely. In seeing impermanence clearly, the grip of craving begins to loosen.

Another key practice is detachment from thoughts. Instead of believing every thought to be true, we can see them as reflections passing across the surface of the mirror. This does not mean ignoring thoughts or pushing them away, but observing them with distance: "This is a thought, not me." Over time, this shift allows us to experience thoughts and emotions without being consumed by them.

Breaking Misconceptions of Self

Through these practices, deep misconceptions about the self begin to dissolve. The first is the idea that "I am my emotions." When the mirror is polished, we see that emotions are temporary visitors, not the permanent structure of who we are. Anger, sadness, or joy may arise, but they do not define the self. Recognizing this creates the freedom to feel fully without being enslaved by feelings.

The second misconception is "I am separate." Buddhist reflection reveals the interdependence, the truth that nothing exists in isolation. Just as a mirror reflects both the subject and the space around it, our lives are reflections of countless causes and conditions: family, culture, food, air, and even the earth itself. To see clearly is to recognize connection, not separation.

The most profound teaching is anatta, or no-self. At first, this idea can sound unsettling: if there is no permanent self, then who am I? However, Buddhism teaches that the self we cling to, with a fixed identity of thoughts, roles, and labels, is an illusion. The mirror of the mind is clear when it reflects experience without needing to grasp an identity from it. This does not erase individuality but frees us from being trapped by it.

Liberation Through Reflection

When the mirror of the mind is free from distortion, what remains is liberation. In Buddhism, this state is known as nirvana, which refers to the extinguishing of craving, ignorance, and attachment. It is not annihilation but clarity: the mirror free of mud, able to reflect life as it is, without distortion.

From this clarity arises compassion. When we no longer see ourselves as separate, empathy flows naturally. Suffering in others is not distant but shared. Just as a clear mirror reflects light evenly, a clear mind reflects kindness and care toward all beings. Compassion becomes less of a duty and more of a spontaneous expression of clarity.

Finally, living with awareness becomes a form of freedom. We are no longer dragged by illusions, no longer chained by distorted reflections. Instead, we move through life awake, calm, and compassionate. Awareness allows us to respond rather than react, to choose peace even in the midst of chaos, and to live with the understanding that the mirror has always been clear beneath the clouds.

Reflection Tools

1. The One-Minute Breath Pause

Stop what you're doing.

Take one slow breath in and one slow breath out.

Notice the sensations of breathing, the rise and fall, the air moving in and out.

Remind yourself: "This is a passing moment, not the whole of me."

Even one minute of mindfulness can polish the mirror when the mind feels clouded.

2. Thought Labeling Practice

When a thought arises, instead of following it, silently label it:

"Planning."

"Remembering."

"Worrying."

"Judging."

By naming thoughts, you step back and see them as passing reflections, not as your identity. This reduces over-identification with the mind's chatter.

3. The Impermanence Journal

At the end of each day, write down:

One pleasant experience that faded.

One unpleasant experience that changed or eased.

One neutral moment that became meaningful when you noticed it.

This practice trains the mind to see impermanence everywhere, loosening craving and fear.

4. Compassion Reflection

Take a few minutes to picture someone you know who is struggling. With each breath, silently repeat:

"May you be free from suffering."

"May you find peace."

"May you be well."

This practice connects clarity with compassion, reminding you that polishing your own mirror allows you to reflect kindness into the world.

5. Mirror Visualization

Close your eyes and imagine your mind as a mirror. Picture smudges of stress, anger, or craving on its surface. With each exhale, gently wipe one smudge away until the mirror shines clear again. End by sitting quietly in the spacious clarity.

Key Takeaway

You don't have to fight thoughts or emotions; you only need to recognize them for what they are. By polishing the mirror daily with mindfulness and compassion, clarity and freedom naturally emerge.

Chapter 2 – Hawaii: Making Things Right

Queen Lili'uokalani's Reflection in Captivity

In January of 1895, Queen Lili'uokalani, the last reigning monarch of Hawaii, was placed under house arrest in her own palace. The overthrow of the Hawaiian Kingdom by foreign interests had already stripped her of power, but when supporters attempted to restore her throne, she was accused of treason. For nearly eight months, she lived in a single upstairs room of 'Iolani Palace, guarded and cut off from her people.

It was a moment that could have hardened her heart. She had been betrayed, humiliated, and imprisoned in her own home. She had every reason to cling to anger. Yet history records that during this time, she turned not to resentment but to reflection.

Each day, she prayed. She wrote in her diary, not only of her sorrow but also of her choice to forgive those who had wronged her. She composed songs filled with themes of faith, reconciliation, and aloha. Her most famous song, "Ke Aloha O Ka Haku" (The Queen's Prayer), written during her imprisonment, is a plea for forgiveness, love, and mercy.

The words echo the heart of ho'oponopono. Though she was cut off from formal ceremony, her practice carried the same spirit: acknowledging pain, reflecting honestly, and releasing resentment so that harmony could be restored. She could not change what had happened, but she could choose how she carried it within herself.

Her reflection was not just personal; it was collective. As a queen, she bore the sorrow of her people as well as her own. If she had hardened her heart, that bitterness could have deepened the wounds of her nation. Instead, she modeled the Hawaiian value of ho'oponopono: "to make right." She sought to preserve her people's dignity, to demonstrate that even in defeat, forgiveness and love could guide the way forward.

After her release, she continued to advocate for her people with grace and persistence. She worked to protect Hawaiian culture and traditions in the face of overwhelming change. And though sovereignty was never restored, she never stopped embodying the principle that reconciliation and forgiveness could preserve unity even when justice seemed out of reach.

Queen Lili'uokalani's story is a reminder that ho'oponopono is more than a family practice; it is a way of being. It is about facing the deepest wrongs not with vengeance, but with reflection, compassion, and a willingness to release bitterness. Her life shows that even in the harshest circumstances, the mirror of the mind can be cleared, and a path of harmony chosen.

Queen Liliʻuokalani's time in captivity is remembered not just as a moment of political upheaval, but as a moment of deep spiritual strength. Stripped of her throne and imprisoned in her own palace, she could have chosen bitterness. Instead, she chose reflection, prayer, and forgiveness. Her words and songs from that period reveal a spirit of reconciliation, a refusal to let resentment rule her heart. This was more than personal resilience; it embodied a Hawaiian principle known as hoʻoponopono, meaning "to make right." The Queen's story reminds us that hoʻoponopono is not only a family ritual, but a way of living. It is about clearing the inner mirror so harmony can be restored, both within ourselves and among those around us.

The Meaning of Hoʻoponopono

The Hawaiian word hoʻoponopono comes from hoʻo (to cause, to make) and pono (right, just, balanced). To say it twice pono pono, is to emphasize complete harmony and correctness. At its core, it means "to make right, to restore balance." Traditionally, hoʻoponopono was more than a personal reflection practice. It was a social and spiritual ceremony used within families and communities to resolve conflict, heal grievances, and restore unity. When tensions grew between siblings over land, between spouses, or even across extended families, a respected elder or kahuna (spiritual leader) would guide the process. The goal was not punishment or blame, but cleansing. Resentment was seen as poison, and to hold it too long was believed to create imbalance, even illness. Reflection was the foundation. Each person was expected to look honestly at their actions, words, and attitudes. By bringing hidden resentments to the surface, the mirror of the community could be cleared. This practice recognized a truth still vital today: conflict left unspoken festers, but conflict faced with honesty can be transformed into connection.

The Ceremony of Reflection

A hoʻoponopono ceremony often began with prayer, inviting divine presence and guidance into the circle. The family or group gathered together, sitting openly with the intention of clearing what was heavy. Silence would be held, then grievances brought forward one by one. Unlike arguments where voices rise and emotions flare, hoʻoponopono was structured and sacred. Elders or leaders kept the tone steady, ensuring that each voice was heard. Words were spoken directly, no hidden whispers or gossip, but honest reflection shared before all. Confession was key. Each person acknowledged their role in the conflict. Whether through harsh words, neglect, or misunderstanding, everyone was called to take responsibility. Through this honesty, the weight of blame was lifted. The process was less about proving who was right and more about dissolving the cloud of hurt.

Forgiveness followed. One by one, family members expressed their willingness to let go. Resentment, like debris on the surface of water, was skimmed away so the reflection beneath could shine clear again. Tears often flowed, but they were cleansing tears, signs of release, not defeat. The ceremony ended with a closing prayer, sealing the restoration of harmony.

The Power of Forgiveness

To the Hawaiian mind, forgiveness was not an optional kindness; it was necessary for balance. Holding resentment was seen as carrying a stone in the heart, one that weighed down both the individual and the group.

Letting go of resentment cleared the inner mirror. A person no longer saw others through the fog of past wounds. They could meet each new moment without dragging yesterday's shadows into it. Trust, once broken, could begin to mend. Forgiveness restored the connection. Families divided by bitterness could return to eating, working, and living together in peace. Neighbors could set aside their hostility and return to cooperation. Forgiveness was understood as more than an emotional act; it was a practical necessity for survival and harmony on the islands.

Modern psychology confirms what Hawaiians long practiced. Studies show that forgiveness lowers stress, reduces depression, and even improves physical health by easing blood pressure and immune strain. Socially, it rebuilds bonds, encourages cooperation, and dissolves cycles of retaliation. In every way,

forgiveness is a medicine for both mind
and community.

Lessons for Today

Though few today participate in
traditional ho'oponopono ceremonies, the
wisdom remains timeless. Families,
workplaces, and even nations can learn
from its principles.

First, reflection is essential for healing
relationships. Avoiding conflict may feel
easier in the short term, but silence breeds
distance. By bringing tensions into the
open with respect, honesty, and
compassion, we make space for release.

Second, forgiveness is not weakness but
strength. It does not erase wrongdoing,
but it frees us from carrying it. Like
polishing a mirror, forgiveness removes
the smudges that keep us from seeing
clearly.

Third, ho'oponopono offers guidance for
modern conflict resolution. Whether in
family therapy, community dialogue, or
even political negotiation, the process of

structured reflection, accountability, and forgiveness can break cycles of hostility. Finally, ho'oponopono teaches us that the past does not have to dictate the future. By reflecting honestly on what has been, we can clear the path for what will be. Balance comes not from pretending conflict doesn't exist, but from facing it with courage, compassion, and the will to make things right.

Reflection Tools

1. The Mirror Clearing Journal

Each evening, write down:

One moment that felt heavy, tense, or unresolved.

How you contributed to it (honest self-reflection).

What you can release through forgiveness (of yourself or others).

One phrase of gratitude to close the reflection.

This practice "polishes the mirror" daily, so resentment doesn't accumulate.

2. The Family Circle Conversation

Inspired by traditional ceremonies, try this simplified version with loved ones:

Sit in a circle and agree that the space is safe and nonjudgmental.

Each person shares one tension or hurt from the week, speaking in "I feel" statements rather than blame.

After sharing, the group closes with a word of appreciation or blessing for one another.

Even brief weekly circles can release misunderstandings before they harden into bigger conflicts.

3. The Forgiveness Visualization

Close your eyes, imagine the person or situation that still weighs on you. Picture a heavy stone in your hands, representing the resentment. Then, in your mind, place the stone into a flowing river or the ocean. Watch it carried away, leaving your hands open and light.

This simple visualization can help you release what your mind clings to, returning peace to your body and heart.

4. Reflection Before Resolution

Before you enter a difficult conversation, take 2 minutes to reflect on these questions:

What am I holding onto that clouds my view?

How do I want the relationship to feel after this talk?

Am I willing to release blame to move toward balance?

Approaching dialogue with reflection first can shift the entire outcome.

Key Takeaway

Hoʻoponopono is not just about saying the right words; it's about courage, honesty, and willingness to clear the mirror of the heart. These practices don't erase conflict, but they transform it into an opportunity for healing and connection.

Closing Thought

Hoʻoponopono reminds us that harmony begins with reflection. Just as Queen Liliʻuokalani chose forgiveness in her darkest hour, we too can choose to clear the inner mirror. When we make right within ourselves and among each other, we make right the world we live in.

Chapter 3 – Greece: Know Thyself and the Light Beyond Shadows

Socrates at His Trial

The year was 399 BCE, and the crowded courtroom of Athens was alive with tension. At the center stood Socrates, a man who had spent his life asking questions that few wanted to answer. His accusers claimed he was corrupting the youth, challenging the city's traditions, and dishonoring the gods. The penalty, if found guilty, would be death.

Most men in his position would have begged for mercy, pleaded with the jury, or at least promised to stop the behavior that had angered so many. Socrates did none of these things. Instead, he stood before the Athenians and delivered a defense that was less about saving his life and more about keeping the truth.

He admitted openly that he did not have the answers people thought he should. In fact, his wisdom, he said, came from knowing his ignorance. He told the jury that many of Athens' leaders claimed to know what justice, beauty, and virtue were. Still, when questioned, their definitions fell apart. His role was not to tear down the city, but to awaken it to act as a gadfly, as he called himself, stinging the great horse of Athens so it would not fall into complacency.

Then came the words that would echo through history: "The unexamined life is not worth living."

For Socrates, this was not rhetoric. He truly believed that to live without reflection (to follow habit, tradition, or desire blindly) was to miss the essence of being human. Life was not about gathering wealth, winning honor, or pleasing the crowd. It was about turning inward, questioning assumptions, and striving for truth. To him, this was the only path to wisdom, and wisdom was the only accurate measure of a good life.

The jury was unmoved. By a narrow margin, they found him guilty and sentenced him to death. Friends urged him to escape; others begged him to apologize or recant his teachings. But Socrates refused. To him, betraying his principles would be worse than dying. If reflection and truth were worth living for, then they were also worth dying for.

In his final hours, he drank the cup of hemlock with calm acceptance. His body may have perished, but his words and example did not. Through his trial, Socrates imparted a lesson to the world that continues to challenge us: that self-knowledge is not optional, that unexamined beliefs lead to distorted lives, and that reflection is the foundation of wisdom.

Socrates' trial showed the world what it meant to stand firmly in the light of reflection, even when surrounded by the shadows of misunderstanding and fear. His words, "The unexamined life is not worth living," were not meant to be lofty philosophy for the few but a challenge for all of us. They remind us that wisdom begins not with outward success, but with turning inward. The Greeks believed that one could only live freely, wisely, and fully by knowing oneself.

The Maxim of Self-Knowledge

Long before Socrates' trial, a message was carved into stone at the Temple of Apollo in Delphi: gnōthi seauton "Know thyself." Visitors came to the Oracle at Delphi seeking guidance about the future, but the first lesson they encountered was that the true key to destiny lay within. To the Greeks, this simple phrase was more than advice; it was a commandment for living well.

Why did self-knowledge matter so much? The Greeks saw reflection as the foundation of wisdom because ignorance of oneself led inevitably to folly. A person who didn't understand their own desires, fears, or limitations was at the mercy of them. Without reflection, life became like wandering through shadows, mistaking fleeting impressions for reality, chasing illusions, and repeating the same mistakes.

Self-deception was the greatest danger. To believe you are wise when you are not, to live by stories that distort reality, or to project blame outward instead of looking inward, these, the Greeks believed, were the roots of tragedy. Their myths often illustrated this danger: hubris, the arrogance of thinking one knows all, nearly always ended in downfall. For them, "Know thyself" was not simply advice for philosophers but a shield against disaster.

Socratic Reflection

Socrates embodied this maxim in daily life. He believed that people often lived according to unexamined assumptions and beliefs inherited from family, culture, or habit. His method was simple yet revolutionary: ask questions until the surface certainty broke apart, revealing either truth or contradiction.

When Socrates asked politicians, poets, and craftsmen what justice, courage, or love truly meant, their confident answers quickly unraveled under scrutiny. What remained was humility, the recognition that certainty was often illusion. For Socrates, this humility was the beginning of wisdom.

That is why, at his trial, he declared: "The unexamined life is not worth living." He wasn't dismissing the value of ordinary life but emphasizing that without reflection, life loses its depth. To live without examining our beliefs, motives, and actions is to live asleep, unaware of the real possibilities of growth.

Socratic dialogue is more than a debate tactic; it is a mirror for the mind. By questioning assumptions, we begin to see the distortions in our thinking. We notice where fear colors our judgments, where pride blinds us, or where stories from the past still dictate our choices. In this way, reflection becomes liberation, a release from the shadows of self-deception.

Plato's Allegory of the Cave

Plato, Socrates' student, carried his teacher's vision further through one of the most powerful metaphors in philosophy: the Allegory of the Cave.

He imagined a group of prisoners chained in darkness, their only view of the world a wall onto which shadows were cast by flickering firelight. To them, the shadows were reality, the whole of existence. If one prisoner were freed and dragged toward the light, the journey would be painful. The brightness would blind him at first, and the unfamiliar shapes would confuse him. Yet slowly, his eyes would adjust. He would see objects, people, landscapes, and finally the sun itself.

If he returned to the cave to tell the others, they might not believe him. They could even resist or ridicule the idea that their shadows were not the truth. But for the one who had seen the light, there could be no going back. Reflection had shown him the difference between illusion and reality.

Plato's cave reminds us that many of our own "truths" may be shadows, distorted reflections shaped by culture, fear, or habit. Reflection is the act of turning toward the light, even when it hurts, because clarity and freedom lie beyond the shadows.

Lessons for Modern Life

The wisdom of ancient Greece still holds power today. In many ways, we live in modern caves surrounded by media, social pressures, and cultural stories that cast shadows on the walls of our perception. Reflection is the way we turn around, step into the light, and see more clearly.

We can practice Socratic questioning in our daily decisions:

Why do I want this?

What assumption am I making?

Is this belief actually true, or have I just accepted it?

These questions, simple as they are, break through automatic thinking. They create space for choice instead of compulsion. Escaping modern caves requires courage. It means recognizing where we are trapped by consumer culture, comparison, or fear of judgment. Just as the cave prisoners resisted the light, we may resist the discomfort of reflection. But clarity is worth the effort, because it reveals what is real beneath the shadows.

Reflection, as the Greeks saw it, is not abstract philosophy; it is a path to wisdom, freedom, and truth. To know yourself is to live awake. It is to recognize the distortions of the mind, to step beyond them, and to live with the clarity that Socrates believed was the essence of a life worth living.

Reflection Tools

1. Socratic Self-Questioning

When facing a decision or strong emotion, pause and ask:

What am I assuming here?

Do I know this is true, or am I guessing?

If I look deeper, what else could be possible?

Write down your answers. This process mirrors Socrates' method of revealing hidden beliefs and testing whether they actually stand.

2. The Cave Check

At the end of the day, reflect:

What "shadows" did I mistake for reality today? (e.g., social media images, a colleague's tone, an anxious thought).

What might the "light" be behind them? (truth, context, compassion).

This helps train the mind to distinguish between distorted perceptions and deeper reality.

3. Know Thyself Journal Prompt

Each week, answer three questions in writing:

What am I most proud of this week?
Where did I react automatically, and what was really going on beneath that reaction?
What truth about myself became clearer?
This builds self-knowledge over time, one layer at a time.

4. Reflection Partner Dialogue

Choose a trusted friend or family member and practice Socratic-style dialogue once a week:

Each person shares a belief or assumption they've been living by.

The other gently asks questions, not to attack, but to help uncover clarity: Why do you believe that? What would it mean if the opposite were true?

This exercise creates a connection while sharpening self-awareness.

Key Takeaway

The Greeks taught that wisdom begins with reflection, not certainty. These practices don't demand you have all the answers, only the courage to question your assumptions and step closer to the truth.

Closing Thought

The legacy of Greece is not only its art, architecture, or politics, but its insistence that the greatest journey is inward. The words carved at Delphi still speak to us across millennia: Know thyself. In doing so, we step out of the cave of illusion, into the light of clarity, compassion, and truth.

Chapter 4 – Hindu and Yogic Traditions: Seeing Beyond the Ego

Yajnavalkya and the True Self

Long ago, in the time when sages wandered the forests of India seeking truth, there lived a wise teacher named Yajnavalkya. Known for his profound insight into the nature of the Self, he devoted his life to teaching students, engaging in debates, and contemplating the mysteries of existence.

One day, Yajnavalkya decided to leave behind his worldly possessions and devote himself fully to the pursuit of spiritual liberation. Before he departed, he called his two wives, Maitreyi and Katyayani, to speak with them. He offered to divide his wealth between them, the cattle, the gold, the household goods, so that they could live comfortably in his absence.

Katyayani, content with material security, accepted without question. However, Maitreyi, renowned for her wisdom and curiosity, was not satisfied. She asked him:

"My lord, if I were to possess the entire earth filled with riches, would it make me immortal? Would it free me from death?"

Yajnavalkya shook his head gently. "No, Maitreyi. Wealth may provide comfort, but it cannot bring immortality."

Maitreyi looked at him intently. "Then what use is wealth if it cannot lead me to the ultimate truth? Teach me instead about that which never dies. Teach me what you know of the Self."

Pleased by her sincerity, Yajnavalkya began to share a teaching that would echo through the ages, preserved in the Brihadaranyaka Upanishad. He told her:

"It is not for the sake of the husband that the husband is loved, but for the sake of the Self.

It is not for the sake of the wife that the wife is loved, but for the sake of the Self. It is not for the sake of wealth that wealth is loved, but for the sake of the Self. It is not for the sake of the gods, or the world, or even knowledge itself that these are loved but for the sake of the Self."

Maitreyi listened, her heart absorbing the depth of his words. Yajnavalkya explained that everything we cling to (relationships, possessions, accomplishments) hinges only because of the Self that experiences them. Without the Self, they hold no meaning. The ego mistakes the reflections in the mirror, status, beauty, and success for reality, but these are only passing forms. The true Self, the Atman, is eternal, unchanging, and pure consciousness itself.

He went on: "The Self is not found by wealth, nor by clinging, nor by craving. It is revealed by reflection, by stripping away the illusions of the ego. Just as the

restless mind is like a rippling mirror,
unable to reflect clearly, so the restless
ego hides the truth of the Self. But when
the mirror is calm, the eternal is seen."
Maitreyi realized that the pursuit of
possessions or praise was like chasing
shadows. True knowledge came not from
acquiring more, but from looking within,
from seeing beyond the ego to the Self
that was always present.

In this exchange, Yajnavalkya gave her
something far greater than wealth: the key
to liberation. The teaching became a
timeless reminder that reflection and self-
inquiry, or svadhyaya, are not about
building up the ego but dissolving it, until
the eternal Self shines through.

The story of Yajnavalkya and Maitreyi
reveals a central truth of Hindu and yogic
philosophy: the ultimate goal of life is not
the accumulation of wealth or status, but
the realization of the Self. This Self, the
Atman, is not the ego that clings to
possessions or roles, but the eternal

consciousness that shines behind all experience. In yogic tradition, the path to this realization begins with reflection. Through self-study, meditation, and mantra, the restless mirror of the mind is polished, allowing us to see beyond the distortions of ego into the wholeness of the true Self.

The Practice of Svadhyaya (Self-Study)

In the eight-limbed path of yoga described by Patanjali, svadhyaya, self-study, is a foundational practice. It is the discipline of turning the gaze inward, examining thoughts, habits, and motives to uncover what lies beneath the surface. Where the ego clings to external validation, svadhyaya asks: Who am I beneath these roles?

Reflection in yoga is not about judgment but awareness. When a practitioner notices impatience arising during a difficult pose, that moment becomes a mirror. What does this impatience reveal about how they approach challenges off the mat? When a craving for praise appears, it becomes a chance to see how the ego longs for recognition. Each observation is an invitation to peel back the layers of illusion.

Sacred texts also serve as mirrors for self-inquiry. The Bhagavad Gita, the Upanishads, and other yogic writings are not meant to be studied like history books but reflected upon as guides for the inner journey. As the Gita teaches, action itself is not the problem; it is attachment to the fruits of action that binds us. To read and reflect on such passages is to hold up a mirror to our own attachments and ask: Where am I seeking permanence in what is passing?

Daily practices of svadhyaya can be simple yet profound. Journaling after meditation, reflecting on a verse from a text, or pausing during the day to ask, "What story is my ego telling me right now?" These small acts gradually polish the mirror. Over time, they reveal that

beneath the chatter of the ego lies
something more profound and
unchanging.

The Upanishadic Vision of the Self

The Upanishads describe the Self as pure
consciousness, eternal, and beyond the
grasp of the ego. While the ego thrives on
separateness and identity, Atman is
untouched, like the sky behind the clouds.
The metaphor of the mirror is used again
and again in Hindu philosophy. When the
mind is restless, clouded by illusion
(maya), the reflection is distorted. We
mistake the passing forms of wealth,
success, and failure for reality. We
identify with labels, believing "I am this
body" or "I am this role." Yet these are
only ripples across the surface.

When the mirror becomes calm, it reflects
the eternal Self clearly. Meditation,
discipline, and reflection settle the surface
of the mind until the Atman shines
through. In that clarity, a person realizes
that the true Self is not the ego but the

same consciousness that animates all beings. This recognition is liberation, the freedom of knowing that what is eternal cannot be harmed or diminished.

Meditation and Mantra as Tools of Reflection

Yoga offers concrete practices to polish the mirror and move beyond the ego. Stillness and breath are the first tools. When the breath is observed with gentle awareness, the restless movements of the mind begin to settle. Just as ripples fade when the wind dies down, so too does stillness reveal the calm beneath thought. Breath becomes an anchor, a reminder that awareness is always here, steady and present.

Mantra is another tool. Yogic traditions teach that sound carries vibration, and sacred words like Om or longer mantras act as tuning forks for the mind. When repeated with devotion, a mantra clears distortions, focusing attention and dissolving scattered thoughts. It is like polishing the mirror with rhythm and resonance until its clarity shines.

Deep meditation is the culmination. In states of profound absorption, the boundary between observer and observed dissolves. The ego quiets, and the practitioner experiences glimpses of pure awareness. These moments of direct realization remind us that we are not our passing thoughts or roles, but the consciousness in which they arise.

Beyond Ego into Wholeness

The goal of yoga is not to destroy the ego but to see beyond its illusions. The ego is like a costume worn in the play of life, helpful in navigating the world, but not the essence of who we are. Reflection helps us move from false identity to true awareness.

To live beyond ego is to live in union. The very word yoga means union — with the Self, with others, with the divine. When the mirror is clear, we recognize that the same consciousness that animates us animates all life. This recognition brings humility, compassion, and freedom.

For modern seekers, these teachings are deeply practical. In a world that constantly reinforces the ego through comparison, consumption, and competition, practices like svadhyaya, meditation, and mantra offer a way back to wholeness. They remind us that we are more than the reflections we chase. Imagine facing criticism without losing your center, or success without becoming inflated. Imagine relating to others not from insecurity but from recognition of shared being. This is the promise of seeing beyond the ego: a life of clarity, compassion, and union.

Reflection Tools

1. Svadhyaya Journal Prompts

At the end of the day, reflect with one or more of these questions:

What thought or belief guided my choices today? Was it rooted in ego or awareness?

Where did I cling to approval, success, or comfort?

What deeper truth about myself became visible through these reflections?

This practice gradually reveals patterns and loosens the grip of false identity.

2. Breath-Centered Meditation

Sit comfortably and close your eyes.

Focus on the breath as it flows in and out.

Each time your mind wanders, gently return to the breath.

Remind yourself: "I am awareness, not my thoughts."

Even 5–10 minutes a day helps settle the restless mind and glimpse the clarity beneath.

3. Mantra Polishing Practice

Choose a mantra that resonates (e.g., Om, So Hum — "I am That").

Repeat it silently or aloud for a few minutes.

Notice how the vibration steadies the mind and clears scattered thoughts.

Let the sound become like a cloth
polishing the mirror of the mind.

4. Ego Checkin Pause

Before reacting to stress or praise, pause
and ask:

Am I responding from my ego or from my
deeper self?

What would my response look like if I
acted from a place of awareness instead?

This tiny pause shifts you from
identification with ego into connection
with the Self.

5. Shared Being Reflection

Look at someone, a friend, family
member, or even a stranger, and silently
reflect:

"The same consciousness that lives in me
lives in you."

This simple practice cultivates humility
and union, dissolving the illusion of
separateness.

Key Takeaway

Ego thrives on clinging, comparison, and
illusion, but reflection polishes the mirror

until the Self shines through. Yoga offers not escape from life, but union with what is most real and eternal within it.

Hindu and yogic traditions remind us that beneath the restless mirror of the ego lies the Atman, pure, eternal, and whole. By practicing self-inquiry, calming the mind through meditation, and resonating with a mantra, we polish the mirror and see clearly. The journey is not about becoming someone new but remembering who we already are: reflections of the eternal Self.

Chapter 5 – Indigenous North American Traditions: Reflection in Nature and Story

Black Elk's Vision

In the summer of 1872, on the Great Plains of what is now South Dakota, a boy lay gravely ill. His family feared he would not survive. At only nine years old, Black Elk, of the Oglala Lakota, slipped into what seemed like a dream but was far more than ordinary sleep.

As his body lay weak, his spirit was carried into a vision that would shape the rest of his life. He later described being taken to the "center of the world," where he was shown the sacred hoop of his people, a great circle symbolizing unity, life, and balance. He saw the four directions represented, each carrying its own wisdom. He was given symbols of power and healing, not for himself alone, but for his people.

In this vision, he saw horses thundering, thunder beings speaking, and the flowering tree of life standing tall at the heart of the world. The message was clear: his role was not merely to live for himself, but to reflect, learn, and return with wisdom that could help restore harmony to his community.

When Black Elk awoke, he was changed. His vision had not been for his own comfort, but for the health of his people. Later in life, as he became a healer and holy man, he would share this story, not as a private possession, but as a collective reflection on the meaning of life. His words became part of the oral tradition, a mirror in which his community could see both their struggles and their strength. Through storytelling, his vision was preserved and passed on. It was not entertainment, but medicine. It reminded listeners that inner reflection and vision were not isolated acts; they were inseparable from the well-being of the tribe, the land, and the generations to come.

Black Elk's vision shows how Indigenous traditions viewed reflection: not as navel-gazing, but as a bridge between the individual and the whole. The time alone, the silence, and the openness to vision were not ends in themselves. They were tools to clear the inner mirror so that wisdom could flow back into the community.

His life became a living example of the belief that harmony within sustains harmony without. By reflecting on his vision and embodying its lessons, he helped guide his people through times of immense change and hardship.

Even today, his story stands as a reminder: reflection is not always about self-improvement in the narrow sense. Sometimes it is about receiving insight that can heal divisions, restore balance, and reconnect us with what matters most, our shared humanity and our relationship with the natural world.

In Indigenous North American traditions, reflection is deeply woven into community, land, and story. Wisdom is not something kept private; it is shared, spoken, and lived together. Storytelling, solitude in nature, and harmony with the community all served as mirrors for ways of seeing oneself more clearly by seeing how life is connected.

Storytelling as Collective Reflection

Indigenous cultures across North America relied on oral tradition as a way of transmitting knowledge, values, and identity. These stories were not written down but carried in memory, passed from elders to the young, shaped and reshaped by each generation.

Oral traditions acted as mirrors of cultural values. A story about a coyote, bear, or raven was never just about an animal; it reflected lessons about humility, courage, generosity, or balance. By listening, individuals could recognize their own struggles and responsibilities. Storytelling made reflection communal; the whole community held up the mirror together. Stories also served as tools for establishing identity and fostering a sense of belonging. A child growing up hearing tales of their ancestors or of the land around them learned not only who they were but where they belonged. Reflection in this sense was not isolated introspection but seeing oneself as part of a lineage, a people, and a world alive with meaning.

Through stories, wisdom was preserved across generations. The lessons of survival, respect for the earth, and the importance of harmony could be carried forward even through hardship. A story told around the fire was more than entertainment; it was a bridge between past and future, a reflective thread binding generations together.

Vision Quests and Solitude

Beyond storytelling, many Indigenous traditions included the practice of the vision quest, a time of solitude in nature, often marked by fasting and prayer. This was not a casual retreat but a sacred rite, often undertaken at key moments of life such as the transition to adulthood.

The practice involved leaving the village and venturing alone into the wilderness, whether to the mountains, the forest, or the plains. Without food, company, or distraction, the seeker was left face-to-face with themselves and the natural world. In that silence, the inner mirror began to clear.

Solitude in the wilderness acted as a teacher. Every sound, every change in the weather, every stirring of the body became a mirror for reflection. Hunger tested endurance. Silence revealed fears and desires. In stripping away the noise of daily life, seekers could see more clearly what lived in their own hearts.

Personal visions often came in dreams, encounters with animals, or sudden moments of clarity. These visions were not random but were understood as reflections of purpose; guidance from the spirit world, the land, or the deeper Self. When the seeker returned to the community, they shared their vision, and it became part of the collective mirror.

Harmony Within, Harmony Without

A central belief in many Indigenous traditions is that inner balance sustains community health. A person who carried unresolved anger or resentment could disrupt harmony, while a person who cultivated clarity and balance could strengthen it. Reflection was not private self-care but a responsibility to the community.

Reflection was also a way of restoring relational harmony. When conflicts arose, the process often involved council, storytelling, and rituals that encouraged individuals to reflect on their actions and intentions. Just as ho'oponopono in Hawaiian tradition cleared resentment, Indigenous North American traditions emphasized the need to speak openly, listen deeply, and restore balance through reflection.

Nature itself was understood as a partner in reflection, not a backdrop. Rivers, mountains, animals, and seasons all carried lessons, serving as mirrors of human experience. A rushing river might reflect restlessness; a still lake might reflect calm. By living close to the land, Indigenous peoples could see their own lives mirrored in the cycles of nature.

Lessons for Modern Life

The practices of storytelling, solitude, and harmony hold deep lessons for today's world.

Reconnecting with nature can serve as a powerful reflective practice. Modern life often separates us from the land, replacing silence with constant stimulation. Yet even short walks in the woods, sitting by a river, or spending an afternoon in quiet outdoor reflection can polish the mirror of the mind. Nature still speaks, if we are willing to listen.

Reviving storytelling is another pathway. Families and communities can share personal stories, cultural histories, or even simple reflections around the dinner table. In doing so, they bring back the communal aspect of reflection, reminding each other of shared values and lessons. Stories heal isolation and restore belonging.

Finally, solitude remains vital for clarity in a distracted world. Phones, screens, and endless tasks cloud the mirror just as surely as cravings and attachments do. Setting aside time for quiet, whether a weekend retreat, a daily moment of silence, or even a few hours alone in nature, can reveal insights that constant activity hides.

The Indigenous traditions remind us that reflection is not only about personal peace but about sustaining balance in families, communities, and with the earth itself. A clear mirror does not just serve the individual; it reflects back harmony to the whole.

Reflection Tools

1. Storytelling Circle Practice

Gather a small group of friends, family, or colleagues.

Choose a simple theme (e.g., "a lesson I learned this week" or "a challenge I overcame").

Each person shares their story without interruption. Listeners only reflect back what they heard, without judgment or advice.

Notice how shared stories create connection and remind us that wisdom is often hidden in lived experience.

2. Nature Sit Spot

Find a single spot outdoors, a tree, a bench, or a quiet corner of a park.

Visit this same spot regularly (daily or weekly) and simply sit for 10–20 minutes.

Observe without agenda: the weather, the sounds, the shifts in your own thoughts. Over time, the spot becomes a mirror in which you'll notice your inner state reflected in how you perceive the environment.

3. Mini Vision Quest (Adapted)

Set aside half a day or a full day to be in nature alone.

Leave behind devices, food, and distractions (bring water for safety).

Spend this time in silence, walking or sitting, noticing what arises in your mind and heart.

At the end, write down what insights or "visions" came through — not as mystical prophecy but as personal reflection from deep stillness.

4. Harmony Journal

At the end of each day, ask yourself three questions:

Did I live in balance today (with myself, others, and the environment)?

Where did I feel harmony?

Where did I feel disharmony, and what might restore it?

This daily reflection helps you track how inner balance shapes your outer relationships.

Key Takeaway

From Black Elk's vision to the countless stories and practices carried through Indigenous traditions, the message is consistent: reflection is not meant to be hoarded but shared. Storytelling, vision quests, and living in harmony with nature all polish the mirror, revealing truths that sustain not only individuals but entire communities as well. In reconnecting with these practices, modern life can find a path back to clarity, belonging, and balance.

Chapter 6 – Taoism: Still Water, Clear Reflection

The Farmer's Fortune

In a quiet village, tucked between mountains and streams, there lived an old farmer known for his calm nature. His days were simple: tending fields, mending fences, and caring for his one beloved horse.

One morning, the horse escaped. The neighbors gathered, shaking their heads in sympathy.

"This is terrible," they said. "Without your horse, how will you plow your fields? What bad luck."

The farmer, brushing the dirt from his hands, replied only:

"Maybe."

Days later, to everyone's surprise, the horse returned, and not alone. It brought with it a small herd of wild horses, strong and untamed. The neighbors crowded around, astonished.

"This is wonderful!" they exclaimed. "Now you have more horses than anyone else in the village. What good fortune!"

The farmer, watching the animals graze, simply said:

"Maybe."

Soon after, his only son tried to ride one of the wild horses. Thrown violently to the ground, the boy broke his leg. Neighbors rushed over again, clucking their tongues.

"How tragic," they said. "Your son is injured, your household weakened. What bad luck."

The farmer, sitting quietly by his son's bedside, answered once more:

"Maybe."

A few weeks later, the army came through the village, conscripting every able-bodied young man for a distant war. Many families wept as their sons were taken away. But the farmer's son, with his broken leg, was left behind.

The neighbors, now in awe, whispered: "How fortunate! Your son was spared. You are blessed with good luck."

The farmer, looking out across his fields, said again:

"Maybe."

The story ends without a final judgment. That is the point. What appears like a disaster may open the door to a blessing, and what appears to be a blessing may contain hidden difficulties. The cycle of events cannot be measured fully in the moment.

In Taoist wisdom, the farmer embodies the sage. He does not cling to "good" or "bad," does not fight against what comes, and does not grasp for certainty where there is none. He lives like still water reflecting events as they are, without distortion, without the ripples of judgment.

The neighbors, by contrast, reveal the restless mind. Each event sends them swinging between elation and despair. They want to control the story, to label it, to fix it into something they can understand. But the flow of life is wider than their judgments.

The parable reminds us of something easy to forget in modern times: reflection is not about endless analysis, nor about rushing to judge. It is about pausing, allowing, and seeing with clarity that events unfold as part of a larger flow. Like water bending around stones, the sage trusts that clarity will return when the mind is still.

Laozi's Imagery of the Sage

Laozi described the sage as someone who does not impose personal will on the world but instead reflects it as it is. Just as a mirror shows whatever stands before it without preference, the sage embodies non-distortion. This doesn't mean passivity or indifference. Rather, it means seeing things clearly, free from the fog of personal bias, fear, or craving. When we let go of the need to twist events into "good" or "bad," we open the possibility of responding wisely instead of reacting blindly.

Among Laozi's favorite metaphors was water. Water is soft yet powerful, yielding yet unstoppable. It flows around rocks instead of fighting them, wears down mountains over time, and nourishes all without striving. To Laozi, water was the perfect image of the Tao itself, flexible, clear, and ungrasping. A mind like water can reflect the truth of life because it does not cling or resist.

Stillness in Taoism is not laziness or withdrawal. It is the strength that comes from not being pulled into every disturbance. Like a pond that clears once the wind stops stirring it, the mind becomes clear when still. From that clarity, strength arises to respond wisely, to adapt, and to flow with life rather than against it. In a world that equates busyness with importance, Taoist stillness reveals a different kind of power: one rooted in clarity and alignment.

Mirrors in the Tao Te Ching

The Tao Te Ching often emphasizes the importance of letting go of desires, rigid plans, and control. This practice of non-attachment can be understood as polishing the mirror of the mind. When we cling to outcomes, judgments, or identities, the mirror becomes clouded. By releasing these attachments, we restore their natural clarity. Just as dust dulls a mirror until it is wiped clean, mental grasping obscures our ability to see reality clearly.

The sage reflects reality without adding distortion. In practical terms, this means pausing before rushing to judgment. A setback at work, for example, might feel disastrous, but when seen without clinging, it may reveal a hidden lesson or new path. Taoist reflection teaches us to allow events to reveal themselves over time rather than stamping them immediately as "good" or "bad."

In Taoism, the sage is not a hero who imposes will on the world. Instead, the sage is like a clear mirror through which the Tao itself can act. By staying open, balanced, and humble, the sage becomes a vessel for wisdom greater than the individual self. This is why Laozi praised humility and simplicity: when the ego steps aside, the deeper flow of the Tao can move freely.

NonJudgment and Non-Clinging

Every judgment we make creates ripples on the surface of the mind. If we insist that something is "terrible," we may miss its hidden blessings. If we cling to something as "wonderful," we may be crushed when it changes. Judgment itself clouds perception, pulling us into cycles of craving and aversion.

Taoist practice encourages loosening our grip on outcomes. This doesn't mean we stop caring or stop acting. It means we act without clinging to the results. Like planting seeds in a garden, we tend the soil and water faithfully, but we do not force the flowers to bloom. Freedom comes from doing what is needed in the moment and trusting the larger flow of life to unfold.

The Taoist principle of wu wei is often misunderstood as "doing nothing." In truth, it means acting in alignment with the natural flow, without unnecessary struggle. When the mind is clear, action arises spontaneously and appropriately, without forcing. Just as water flows naturally downhill, so too does the sage move with the grain of life. Reflection allows us to sense where the flow is, so we can join it instead of fighting against it.

Living in Harmony with the Tao

Life is full of storms, both literal and emotional. The Taoist path does not promise calm seas at all times. Instead, it teaches how to keep the inner mirror steady even as the world shifts. This may mean breathing deeply during conflict, pausing before reacting, or returning to silence when overwhelmed. Clarity is not the absence of disturbance, but the ability to see through disturbance.

The Tao Te Ching praises simplicity and humility as ways to stay aligned with the Tao. When we let go of the need to dominate, impress, or accumulate endlessly, we rediscover balance. Reflection helps us notice when we've strayed too far into excess or ambition and gently guides us back to simplicity. In this way, reflection is not abstract but profoundly practical, a compass for daily living.

Ultimately, Taoist reflection is about alignment. The mirror of the mind is not polished so we can admire ourselves, but so we can see the flow of life clearly. When we stop fighting against what is and begin to move with it, life feels less like a battle and more like a dance. We find that reflection is not an escape from life, but rather a doorway to living in harmony with it.

Reflection Tools

1. The "Maybe" Pause

When something happens, good or bad, resist labeling it immediately.

Silently say to yourself: "Maybe."

This creates space to see how the event unfolds over time rather than locking into judgment.

Practice this for a week and notice how it softens your reactions.

2. Water Watching

Find a river, pond, or even a glass of water.

Spend 5–10 minutes observing how water reflects, adapts, and flows.

Ask yourself: "How can I bring this quality into my own life today?"

Use water as a living metaphor to guide your choices.

3. Non-Grasping Journal

Each evening, write down one thing you grasped for that day (a result, an expectation, or an outcome).

Then write how it felt when you let it go, even briefly.

This practice "polishes the mirror" by noticing where clinging clouds your clarity.

4. Wu Wei in Action

Pick one task in your day, cooking, walking, or working, and do it with wu wei, or effortless action.

That means no rushing, no forcing, no multitasking.

Simply move with the natural rhythm of the task, letting it unfold at its own pace.

Notice how calm and clear you feel when you stop pushing.

These practices help readers experience Taoist principles not as abstract ideas but as daily reflections: staying still, flowing like water, and letting go of grasping.

Key Takeaway

The farmer's "Maybe" was not indifference. It was wisdom born of clarity, the recognition that life is too vast to be captured in the snap judgments of the restless mind. Taoism invites us to cultivate that same clarity: to keep the mind still, to flow like water, and to mirror life without distortion. In doing so, we discover a freedom deeper than control and a peace stronger than certainty.

Chapter 7 – African Wisdom Traditions: Reflection Through Ancestors and Ubuntu

The Griot Tradition of West Africa

In many parts of West Africa, history is not confined to the pages of a book; it is carried in the voices of the griots. For centuries, these poet-historians have served as living libraries, passing down stories of families, kingdoms, and communities through song, rhythm, and spoken word. Their role was not entertainment. It was survival.

When a child was born, the griot could recite the lineage, naming ancestors stretching back generations. At weddings, they sang of family bonds and the values that would sustain the new household. At funerals, they reminded the living of the lessons and dignity carried by the deceased, ensuring that memory did not fade into silence.

One of the most famous griot traditions comes from the Mandé peoples of Mali. In the 13th century, griots preserved the story of Sundiata Keita, the founder of the Mali Empire. They told how a boy, once thought too weak to walk, rose to become a great king, uniting the region under justice and prosperity. The griots did not simply praise his achievements; they highlighted the struggles, doubts, and lessons along the way. These stories became moral mirrors, teaching each generation that strength lies not only in power but in resilience and reflection.

The griot's role was communal, not individual. A family might call on a griot to settle disputes, reminding both sides of the values and history that bound them together. A village might seek guidance from a griot during hard times, listening as ancestral wisdom was woven into song. The message was always the same: you are not isolated, you are part of a larger story.

In this way, griots were more than storytellers. They were custodians of identity. They ensured that no one forgot where they came from, that the wisdom of the past remained alive in the present. Through reflection on ancestors, people discovered their place in the community. Through stories of hardship and triumph, they learned that their lives were chapters in an ongoing tale.

This tradition survives today. In Mali, Senegal, Guinea, and Gambia, griots still perform at ceremonies, carrying drums, koras (a 21-string harp), and words rich with memory. In modern times, griots have even taken their art to new stages, from festivals to international tours, sharing not just music but the philosophy of Ubuntu, the recognition that "I am because we are."

Listening to a griot is not passive. The audience reflects as they hear. A person might see their struggles mirrored in the stories of ancestors. A community might recognize patterns from history repeating in the present. Reflection becomes a collective act: the griot offers the mirror, but the people themselves must look into it.

The griot tradition shows us something universal: reflection is never only about the self. It is about identity woven into relationships with family, community, land, and ancestors. The griot reminds people that they are more than individuals striving for private success; they are inheritors of wisdom, participants in a shared story, and contributors to a future that others will one day remember.

In a modern world obsessed with personal achievement, the griot's voice echoes with a different kind of truth: your worth is not just what you achieve alone, but the harmony you create with those around you. This is the essence of reflection through Ubuntu, the recognition that life's

mirror does not stop at the edge of the self
but extends outward into the shared fabric
of humanity.

Ancestors as Living Mirrors

In African wisdom traditions, ancestors
are seen as active participants in the lives
of their descendants. They are mirrors of
continuity, reminding the living of where
they come from and what values hold
them together. Reflection, in this context,
is not just self-examination but a dialogue
across generations. When challenges
arise, people look not only within but also
backward, asking: What did those before
us teach? How can their wisdom light our
path today?

Rituals of remembrance are central to this
worldview. Pouring libations, reciting
lineages, and calling the names of
ancestors are not empty ceremonies. They
are acts of reflection that anchor the
individual in a larger story. By
acknowledging the presence of those who
came before, the living clear the mirror of

forgetfulness and see themselves as part of a continuum. These practices teach that reflection is not only personal clarity but also cultural memory.

For many African communities, identity is inseparable from ancestry. To know oneself fully means to know the line of people, values, and struggles that made one's existence possible. This continuity gives strength, dignity, and resilience. When people remember their ancestors, they also reflect on their own role: What kind of ancestor will I become for the generations after me? Reflection, then, becomes not just about correcting today but about shaping a legacy for tomorrow.

Ubuntu — "I Am Because We Are"

Ubuntu is one of the most widely recognized African philosophies of reflection and community. It teaches that a person's humanity is realized through others: "I am because we are, and since we are, therefore I am." This principle turns the mirror outward. Selfhood is not discovered in isolation, but rather in relationships through generosity, cooperation, and compassion.

Ubuntu invites individuals to reflect not only on themselves but on their connections. It asks: How do my choices affect others? How do others shape who I am? Reflection becomes an act of seeing the web of interdependence clearly. Where the illusion of separateness distorts the mirror, Ubuntu restores accuracy by showing that the self and the community are two sides of the same reflection.

Ubuntu also restores dignity by affirming that every person, no matter their status or mistakes, belongs to the community. Through Ubuntu, reflection is not a harsh judgment but a compassionate reintegration. This principle was famously invoked in South Africa's Truth and Reconciliation Commission, where victims and perpetrators of apartheid shared their stories in public reflection. The process was painful, but rooted in

Ubuntu, the recognition that healing and dignity can only come when people see themselves as part of one human family.

Healing the Illusion of Separation

Modern life often distorts the mirror of the mind by overemphasizing individualism. The message is that worth comes from standing apart, competing, and proving oneself superior. In African wisdom traditions, this is seen as a dangerous distortion of a mirror that fragments rather than unites. When the self is defined only in isolation, alienation and insecurity follow.

African traditions offer a different view: community itself is the truest mirror. To see oneself clearly is to see how one lives with and for others. Just as a face cannot be seen without a mirror, the self cannot be fully seen without community. Reflection in this sense is collective; it requires dialogue, relationships, and a sense of belonging.

Ubuntu heals the illusion of separateness by affirming that every person is part of the whole. A single hand cannot clap by itself; a single tree cannot make a forest. Reflection through Ubuntu reminds people that strength and meaning come from togetherness. Alienation dissolves when individuals recognize that they are not outsiders struggling alone but essential parts of a shared humanity.

The Communal Self as True Reflection

In African wisdom traditions, the communal self is the truest reflection of identity. A person is not diminished by belonging to the group they are completed by it. Reflection in this sense is not about ego but about alignment with community values, harmony, and continuity. The "I" is always within the "we."

Reflection as a collective practice of harmony

Reflection, therefore, is practiced not just in solitude but in circles, rituals, and conversations. Communities gather to tell stories, sing songs, and consult elders because clarity is born collectively. Through these practices, the community polishes its mirror together, ensuring that misunderstandings, resentments, and divisions do not cloud the reflection.

Applying Ubuntu in today's fractured societies

The wisdom of Ubuntu and ancestral reflection speaks directly to the modern world. In societies fractured by division, polarization, and loneliness, the African traditions remind us that healing begins by seeing ourselves in others. Modern reflection, if guided by these principles, could mean listening deeply before judging, honoring the wisdom of elders, and making choices that prioritize the well-being of the whole. Ubuntu challenges us to ask not only, "Who am I?" but also, "Who are we?"

Reflection Tools

1. Ancestral Reflection Journal

Each morning or evening, write down the name of an ancestor, elder, or family member (living or passed) who influenced you.

Ask: "What did they teach me?" and "How can I carry that forward today?"

If you don't know your ancestry, choose cultural figures or role models who represent the values you want to embody.

2. Ubuntu CheckIn

When making a decision, pause and ask:

How does this affect me?

How does this affect others?

How does this affect the whole?

This three-part reflection mirrors the Ubuntu principle: "I am because we are."

3. Storytelling Dinner

Once a week, invite family, friends, or colleagues to share a story at the table. The theme could be "a challenge I overcame," "something I learned from an elder," or "a moment of unexpected kindness."

Listening to stories as mirrors of shared humanity helps dissolve isolation and build connection.

4. The Ancestor's Chair

Place an empty chair in your room and imagine an ancestor or mentor sitting in it. When facing a challenge, ask: "What would you have me see? How would you guide me?"

This visualization makes reflection a dialogue, reminding you that wisdom flows from beyond your immediate perspective.

5. Ubuntu in Action

Choose one daily act that affirms interconnection:

Smile at a stranger.

Call someone you've neglected.

Offer help without expecting anything in return.

Afterward, reflect: "How did this change how I see myself and others?"

These tools bring the spirit of Ubuntu and ancestral reflection into modern life, not as abstract philosophy, but as a lived practice of connection, gratitude, and harmony.

Key Takeaway

African wisdom traditions remind us that reflection is not just about personal clarity, but also about communal health. The ancestors serve as mirrors of continuity, Ubuntu restores the truth of interdependence, and community itself becomes the lens through which selfhood is seen. When reflection is practiced in this way, it heals the illusion of separation and restores belonging. It shows that to live clearly is not only to know oneself but to embrace one's place in the great web of humanity.

Chapter 8 – A Common Thread: The Mirror Across Cultures

The Monk and the Traveler

Long ago, a traveler set out on a journey across the world with one burning question: "How can I truly know myself?" He had read countless books, listened to teachers, and pondered in solitude, yet the answer remained elusive. He believed that perhaps, somewhere in a faraway land, he would find the key.

His first stop was Greece. Standing before the Temple of Apollo at Delphi, he saw words carved in stone: "Know thyself." The priests told him that wisdom begins with reflection, and that ignorance of one's own nature was the greatest danger of all. But how was one to know the self? They gave no clear method, only the challenge: turn inward, question your assumptions, and clarity will follow.

Unsatisfied, the traveler continued east. In India, he encountered yogis who spoke of Atman, the true self beyond the ego and passing thoughts. "The mind," they said, "is like a mirror. When restless, it reflects confusion. When calm, it reflects pure consciousness." They taught him practices of self-study, meditation, and mantra. For the first time, he felt the stirrings of stillness within.

Still seeking, the traveler went on to China, where he found sages gathered along rivers and mountains. They spoke of the Tao and pointed to water as the greatest teacher. "See how it flows without resistance, how it reflects the sky when still, how it nourishes all without striving. Be like water," they said. Here, the traveler understood that clarity was not something to seize but to allow.

From there, he journeyed southward into Africa, where elders welcomed him into their circle. They did not speak of the self in isolation but in relation to others. "Ubuntu," they said. "I am because we are." In stories and rituals, they showed him that identity is a mirror held by the community, through which one cannot see oneself clearly except through the eyes of others. Reflection here was not a private act but a collective one, healing division and restoring dignity.

Finally, he sailed across the ocean to the Hawaiian Islands, where he witnessed a ho'oponopono ceremony. Families gathered, speaking openly of wrongs, asking forgiveness, and releasing resentment. "To reflect on the past," an elder explained, "is to make the future right. Harmony is restored when the mirror of the heart is clear." The traveler realized that reflection was not only for the individual but for the healing of relationships and communities.

When his journey ended, the traveler sat alone beside a quiet stream. He looked into the water, watching how every breeze distorted its surface and how every moment of stillness restored its reflection. He recalled Greece's call to "know thyself," India's teaching of the calm mirror, China's wisdom of water, Africa's Ubuntu, and Hawaii's practice of reconciliation. Different languages, different customs, yet all pointed to the same truth: the mind is a mirror.

He finally understood what had escaped him. Knowing the self was not about discovering a hidden secret somewhere out in the world. It was about polishing the mirror of the mind through questioning, stillness, community, and forgiveness until reality reflected back with clarity. He had traveled the world to learn what had been with him all along.

With a smile, the traveler dipped his hand into the stream, rippling its surface. For a moment, his reflection disappeared. But as the water settled, his face returned clearer, calmer, and somehow freer.

The traveler's journey across lands and traditions revealed something profound: no matter the language, culture, or historical context, people have often turned to the same images to describe the mind and the work of reflection. Mirrors, water, stillness, and clarity appear again and again, suggesting a universal intuition: when the mind is calm, truth is revealed. When agitated, distortion and illusion take over. This shared thread weaves humanity together, showing that reflection is not the exclusive possession of any one culture, but rather part of our common inheritance.

The Shared Metaphor of Reflection

From the Delphic inscription "Know thyself" in Greece to Buddhist descriptions of the mind as a polished mirror, cultures across the world have used the mirror as a metaphor for consciousness. A mirror reflects whatever is placed before it accurately when clean, inaccurately when covered in dust or scratches. Likewise, the mind reflects experiences, emotions, and beliefs. When it is burdened by fear, craving, or confusion, the reflection is distorted. When clear, it shows reality as it is. Water appears just as often in these traditions. Taoism speaks of water as the ultimate model: humble, adaptable, and reflective. Indigenous traditions describe still lakes or flowing rivers as mirrors of the soul. Even in the West, poets and philosophers have looked to water as a symbol of clarity and renewal. Still water reflects the sky, just as a calm mind reflects truth. Disturbed water, like a restless mind, creates ripples that break the image into fragments.

Why do mirror and water metaphors appear everywhere? Because they capture a truth visible to anyone who pauses to look. Every human being has seen their reflection distorted in moving water or made hazy by a dirty mirror. These everyday experiences become metaphors for inner life: agitation distorts, stillness clarifies. The fact that cultures separated by oceans and centuries reached the same imagery suggests reflection is a universal human recognition, grounded in lived experience.

Agitation Distorts, Clarity Reveals

Modern science confirms what ancient wisdom intuited: stress and fear narrow perception, bias the mind toward threats, and create distortions in judgment. In psychological terms, the fight-or-flight response clouds reflection, reducing openness to nuance. Like ripples on water, stress prevents us from seeing clearly. A conversation may be heard as hostile when it is neutral. A challenge may feel like a catastrophe when it is simply an obstacle.

Just as water must be still to reflect the sky, the mind must be calm to reflect reality. Calmness does not mean the absence of movement or challenge; it means balance. When the nervous system is regulated, when breath is steady, perception widens. Possibilities appear. This is why reflection practices across cultures emphasize slowing down, breathing, pausing, or entering silence. Stillness creates the condition for clarity. When the mind is agitated, it generates illusions. A person may believe they are unworthy, unloved, or permanently trapped by circumstance. A community may project fears onto outsiders, creating enemies where none exist. Nations may demonize one another, acting out of distorted reflection rather than clear perception. These illusions are costly: they lead to conflict, isolation, and suffering. Recognizing distortion is the first step toward restoring clarity.

Universal Medicines of Reflection

From Hawaiian ho'oponopono to South Africa's Truth and Reconciliation Commission, reflection has often been formalized into rituals of forgiveness. These practices allow communities to pause, recall the past honestly, and release resentment. Forgiveness clears the mirror, removing the stains of anger and vengeance so trust can return. Across cultures, ceremonies of reconciliation demonstrate that reflection is not merely a personal healing process, but a communal medicine.

Buddhism, Taoism, Hinduism, Christianity, Islam, and Indigenous traditions all include practices of meditation, prayer, or silence. Although the methods differ, the goal remains the same: to calm the mind, polish the mirror, and see more clearly. Whether sitting in vipassana meditation, chanting a mantra, or entering a desert retreat, people throughout history have discovered that stillness itself is a medicine for distortion. Beyond meditation, other reflective practices abound. Indigenous North American traditions use storytelling and vision quests, where solitude and fasting open space for insight. African griots use oral histories to connect individuals to community and ancestors. Across traditions, reflection is not an abstract concept but something woven into lived practices. These "universal medicines" remind us that clarity is accessible through simple, human actions: silence, story, song, and solitude.

Healing the Individual and the Community

Reflection as personal healing of wounds and misconceptions

On an individual level, reflection allows people to see through the illusions created by fear, stress, and old wounds. It helps someone recognize that a passing thought is not their identity, or that a painful memory does not define their worth. Reflection polishes the inner mirror so the self can be seen as whole, capable, and free. Healing begins when a person stops fighting distorted reflections and learns to rest in clarity.

Communities, too, require reflection to heal. Ceremonies of truth-telling, group dialogues, and shared rituals help people acknowledge the past without becoming trapped by it. Reflection allows communities to reestablish trust, to see one another not as enemies but as fellow participants in a shared story. Just as individuals need clear mirrors, so too do groups need collective mirrors to guide them toward harmony.

At the broadest level, reflection is the foundation of societal well-being. Societies that pause to reflect on history, on injustice, on shared humanity are more likely to build systems of fairness and cooperation. Societies that refuse reflection are vulnerable to repeating mistakes, perpetuating conflict, and clinging to distorted self-images. Reflection, therefore, is not just a private virtue but a public necessity.

Key Takeaway

From Greece to Africa, from Taoist rivers to Buddhist mirrors, humanity has circled around the same truth: the mind reflects reality, but only when it is calm and clear. Agitation distorts; clarity reveals. Across cultures, people have discovered medicines to restore the mirror: silence, forgiveness, story, ritual, and stillness. And again and again, they have seen that reflection heals both the individual and the community. The traveler's journey was not just his own; it was humanity's story. The mirror is one, though we look into it from many directions.

Part 2 – The Mirror of the Mind

How the way we reflect on ourselves and the world shapes reality. Ancient wisdom and modern science are converging on this truth.

Throughout history, people have compared the mind to a mirror or water. These metaphors captured an essential truth: clarity allows us to see reality as it is, while agitation distorts. What was once described through story and symbol is now being confirmed by science. Neuroscience, psychology, and contemplative research reveal that reflection literally changes the brain, reshapes perception, and improves well-being. Ancient wisdom and modern findings are converging, pointing toward a shared understanding: the way we reflect on ourselves and the world does not just shape our thoughts, it shapes our reality.

The Scientific Lens on Reflection

Modern neuroscience shows that the brain is not a passive recorder of reality. It actively constructs experience. What we notice, how we interpret, and what meaning we assign to events all depend on mental filters shaped by past experiences, beliefs, and states of mind. Self-awareness, the capacity to "see ourselves seeing," is central to this process.

The Default Mode Network (DMN), a network of brain regions activated during self-reflection, helps us process identity, memory, and future planning. When overactive, it can trap us in cycles of rumination or self-criticism. When balanced, it allows for healthy reflection, integrating lessons from the past and envisioning future possibilities. In essence, neuroscience confirms what wisdom traditions suggested: the "mirror of the mind" shapes not just what we think, but how we live.

Psychologists distinguish between healthy reflection and destructive rumination. Reflection involves curiosity, perspective, and learning, asking "What can I understand here?" Rumination, by contrast, loops on blame, fear, and self-judgment, asking "What's wrong with me?" Both use similar mental processes, but the outcomes differ dramatically. Reflection clarifies; rumination clouds. This distinction echoes the ancient teaching that clarity leads to wisdom, while agitation leads to distortion.

Scientific studies now show that mindfulness, the practice of observing thoughts and emotions without clinging to them, can rewire the brain. Regular practice reduces activity in the DMN, increases regulation in the prefrontal cortex, and strengthens connections to regions linked with empathy and compassion. The result: lower stress, greater emotional balance, and improved overall well-being. Science, in its language of neurons and networks, affirms what traditions long taught in the language of mirrors and water: calm reflection restores clarity.

Ancient Wisdom Meets Modern Findings

Long before psychology became a field of study, traditions worldwide emphasized the importance of reflection. Buddhists described the mind as a mirror to be polished. Taoists taught that stillness was the path to clarity. Indigenous traditions created rituals for communal reflection, showing how honesty and forgiveness heal relationships. African philosophies like Ubuntu emphasized identity through interconnection. What neuroscience calls "self-awareness," these traditions are described in practices of meditation, storytelling, ritual, and self-inquiry. The metaphors of mirror and water are not poetic accidents; they align with how the brain actually works. Just as a dirty or rippled surface distorts an image, a stressed or distracted brain distorts perception. And just as still water reflects clearly, a calm brain shows reality more accurately. Neuroimaging studies of mindfulness reveal that calm states increase perceptual accuracy and emotional regulation, literally allowing people to "see more clearly." Science, in its data, validates what wisdom traditions intuited through lived experience.

Across time and culture, reflection has been linked to flourishing because it enables people to see beyond illusions, whether these are illusions of personal inadequacy, social division, or fear of the unknown. Reflection turns raw experience into meaning. It transforms suffering into growth, conflict into reconciliation, and confusion into clarity. This is why every culture developed its own reflective practices: because without them, the human mirror clouds, and life becomes reactive rather than responsive.

Reflection as a Bridge Between Worlds

Reflection is not an escape from the world but a preparation to engage with it. Neuroscience shows that clarity in the brain improves decision-making and resilience. Wisdom traditions show that a clear mind brings compassion and wisdom to relationships. Both point to the same bridge: inner clarity leads to outer transformation. When the mirror is clear, actions align with values, and life flows with greater harmony.

Social change often begins with personal reflection. History shows that movements for justice and reconciliation were born from moments of deep introspection, as individuals questioned the stories they had been told and communities acknowledged truths they had previously avoided. Science explains this in terms of perspective-taking and empathy; traditions frame it as awakening or returning to balance. Both affirm that the path to collective healing begins with individual reflection.

We live in a moment where ancient wisdom and modern science are converging. The language differs between neurons and networks versus mirrors and water. Still, the message is strikingly similar: the way we reflect shapes the way we live. Reflection is both inner practice and outer necessity, both personal healing and communal restoration. The mirror of the mind stands as a bridge between worlds, science and spirituality, past and future, self and society.

Key Takeaway

Part 3 explores this convergence in depth. We will see how modern neuroscience and psychology illuminate the mechanics of reflection, how ancient traditions anticipated these findings, and how together they offer a vision for the future. Reflection is no longer just a metaphor. It is a science, a practice, and a shared human inheritance. When the mirror of the mind is polished, individuals thrive, communities heal, and societies move toward harmony.

Chapter 9 – How the Brain Reflects: The Science of Self-Awareness

The Neural Mirror Lens

By the late 2040s, personal technology had moved far beyond smartphones and smartwatches. What dominated the public imagination was the Neural Mirror Lens, a wearable headset no thicker than a pair of reading glasses. On the surface, it looked like ordinary eyewear. But behind the clear frame rested a breakthrough: the ability to translate brain activity, especially the Default Mode Network (DMN), into real-time visual feedback. The DMN, long studied as the brain's reflective system, had always been elusive in relation to everyday awareness. People could read about how it fueled daydreams, self-criticism, or creative leaps, but few could see what it was doing moment to moment. The Neural Mirror Lens changed that. For the first time, the private landscape of thought was given a simple, intuitive reflection.

When the mind slipped into worry or rumination, the lens displayed turbulent patterns across the wearer's vision. It was as though storm clouds appeared on glass, rippling and jagged, blocking clarity. If the inner dialogue grew harsh, the distortion intensified: colors dulled, lines bent, and the world outside seemed subtly warped. However, when the wearer returned to a calm awareness through breath, presence, or a shift in attention, the storm receded. The distortion cleared, replaced by luminous clarity, as if sunlight broke through fog.

The impact was immediate and profound. At first, the lens was marketed as a tool for mental health and mindfulness. Therapists used it in sessions to show patients what anxiety looked like in neural terms. A client describing an endless loop of self-criticism could watch their lens fill with chaotic waves, then practice calming breaths until the waves subsided. For the first time, people could literally see the effects of their own choices of attention. Schools quickly adopted it as well. Children prone to distraction could watch the ripples of their wandering minds fade into steadier patterns when they focused. For many students, learning to reflect was no longer abstract; it became a direct, embodied skill. A teacher could say, "Notice the storm? That's not you. That's just your mind swirling. Try letting it settle." And students, empowered by the visual, learned to return again and again to clarity.

As the Neural Mirror Lens spread, society began to change in unexpected ways. In homes, family conflicts softened. Parents and children wearing the lenses could see how their emotions triggered each other. A raised voice in anger didn't just sound harsh; it appeared as jagged spikes of distortion, filling the shared space with turbulence. Arguments became harder to justify when the evidence of harm was literally visible.

In workplaces, leaders started to train with the lens to keep meetings productive. A manager could pause when noticing their own distortions rising, choosing not to project hostility or fear onto neutral situations. Over time, organizations found that conflicts resolved faster and creativity flowed more freely when employees learned to calm the storm before it spread.

Artists and musicians even found inspiration in the device. Some began projecting their Neural Mirror visuals during performances, turning their inner states into public art. A pianist playing Chopin while wearing the lens revealed a stunning dance of clarity and turbulence across a concert hall wall, a moving reminder that behind every note lay the shifting mirror of the mind.

But the invention also raised troubling questions. If the lens could reveal distortions, could it be misused to judge people? Governments and corporations were tempted to demand employees wear it during high-stakes decisions. Some feared the rise of "thought surveillance," where even private mental storms could be monitored and recorded.

Critics argued that true reflection required inner ownership, not technological outsourcing. "If we depend on the lens to tell us when we're distorted," one philosopher warned, "we risk losing the ability to sense it on our own. The mirror must be built inside, not strapped to the face."

And yet, others saw the device as a stepping stone. Just as training wheels help a child learn balance, the Neural Mirror Lens offered a temporary scaffold. The ultimate goal, advocates argued, was not reliance but independence, eventually removing the glasses once the inner mirror had been polished.

By the early 2050s, a movement emerged around what educators called Reflective Literacy. Just as reading and writing had once been taught as universal skills, now reflection was considered essential for a healthy, thriving society. The Neural Mirror Lens became a learning tool, not a permanent crutch. People used it to understand their patterns of thought and emotion, then gradually practiced sensing the distortions without needing the visual aid.

Graduates of this training reported profound changes. Stressful situations no longer hijacked them as easily. Conflicts lost their intensity. Creativity flowed more naturally, unchained from the loops of self-doubt. The lens had not just shown them their minds, it had taught them how to trust their own awareness.

In the end, the Neural Mirror Lens was less about technology and more about reminding humanity of an ancient truth: that the mind is already a mirror. What the device reflected with pixels and overlays had been present all along in the quiet space of awareness.

For some, the lens was revolutionary. For others, it was simply a bridge back to practices humans had always known, sitting in stillness, watching the ripples of thought pass, and waiting until the water cleared. Whether through silicon or silence, the message was the same: clarity is possible, and the reflection can be made whole.

The Power of Metacognition

To be human is to think. But what makes us unique is our ability to step back and notice the fact that we're thinking at all. That's metacognition: the awareness of our own mental processes. It's like turning a mirror inward, catching a glimpse of the workings of the mind instead of being lost inside them.

Think about the last time you caught yourself spiraling in self-doubt. Maybe you heard the thought, "I'm not ready for this," right before a big meeting. Metacognition is what allows you to pause and say, "That's just a thought. It doesn't define me. I can choose how to respond." Without that pause, we're pulled along by the current of whatever shows up in our heads. With it, we step into choice and freedom.

Metacognition supports learning and growth because it helps us evaluate our own mental strategies. Students who reflect on how they study, not just what they study, tend to learn faster. Adults who notice how they handle stress can adjust their habits. Leaders who reflect on their decision-making can improve their judgment over time. This layer of "thinking about thinking" makes change possible because it breaks the automatic loop.

Awareness is always the first step to transformation. If we don't notice the pattern, we can't shift it. A fogged mirror shows only distortion. A clear one shows the truth, and truth is what sets us free.

The Default Mode Network (DMN)

Now let's zoom into the brain itself. Neuroscientists have discovered that when your mind isn't focused on a specific task, it tends to wander into self-reflection. This is thanks to a set of brain regions called the Default Mode Network (DMN).

The DMN is like your brain's built-in reflective system. It helps you think about yourself, imagine the future, remember the past, and consider other people's perspectives. Without it, we wouldn't be able to plan our lives, empathize with others, or even tell coherent stories about who we are.

But like any system, it has a shadow side. When the DMN becomes overactive, it turns reflection into rumination. Instead of helpful self-awareness, we end up replaying regrets, worrying about the future, or criticizing ourselves endlessly. The DMN, when left unchecked, becomes the inner critic's favorite megaphone.

The balance is where the magic happens. A calm and balanced DMN supports creativity, imagination, and problem-solving. It allows us to daydream productively, connect dots in new ways, and envision possibilities. Think of it as the difference between being trapped in a negative thought loop and suddenly having a breakthrough idea while taking a shower. Both moments involve the DMN, but one clouds the mirror while the other clears it.

Benefits of Reflective Awareness

When we train ourselves to notice our thoughts without being swept away, something powerful happens: clarity. We begin to separate reality from distorted thought. Instead of assuming, "They didn't text back, so they must be mad at me," we can pause and ask, "What else might be true?" That pause creates space for truth to emerge.

Clarity also feeds directly into emotional regulation. Instead of being yanked around by every wave of anger, anxiety, or shame, we can learn to watch those emotions rise and fall like storms passing through the sky. The brain literally shifts from reactive regions like the amygdala into more balanced networks when we practice reflective awareness.

And with clarity and regulation comes better decision-making. Reflection helps us step back and see the bigger picture. Rather than reacting from fear or habit, we can choose actions that align with our values. That's the difference between quitting a project at the first setback and sticking with it because you know it matters. Or between snapping at a loved one in frustration and pausing long enough to respond with kindness.

Training the Reflective Brain

The good news is that reflection is trainable. Just like muscles grow with use, the brain's reflective systems strengthen with practice. Mindfulness is one of the most powerful tools here. Research shows that mindfulness meditation literally quiets overactivity in the DMN, reducing rumination and worry. Instead of being trapped in loops, the brain learns to rest in awareness.

Beyond meditation, there are many ways to strengthen metacognition. Journaling, for example, forces us to step back and notice our thoughts on paper. Asking ourselves simple questions, such as "What am I feeling right now? Why might I be reacting this way?" brings unconscious patterns into the light. Even short mindfulness pauses throughout the day help retrain the mind to observe instead of react.

Ultimately, reflection is mental fitness. Just like physical exercise keeps the body strong and adaptable, reflective practices keep the mind clear and resilient. A reflective brain doesn't get stuck in old loops; it adapts, grows, and chooses with awareness. It's not about eliminating thoughts or feelings; it's about seeing them for what they are and remembering that we are more than the reflections passing through.

Reflection Tools

1. Thought-Spotting Pause

Several times a day, stop for 30 seconds.

Ask: "What am I thinking right now?"

Notice the thought without judging or trying to change it.

Say silently: "This is just a thought, not who I am."

Return to what you were doing.

Why it works: Builds metacognitive awareness — you begin to notice thoughts instead of being swept away by them.

2. Name the Storm, Watch It Pass

When a strong emotion arises (anger, anxiety, sadness), pause and name it: "This is anger."

Picture it like a weather pattern moving across the sky.

Breathe slowly for five breaths, then remind yourself: "Like a storm, this will pass."

Why it works: Shifts the brain from emotional reactivity to reflective awareness, calming the Default Mode Network.

3. Reflective Journaling Prompt

At the end of each day, write down:

One thought I noticed today that wasn't helpful.

What else might be true besides that thought?

How did my choices change when I stepped back?

Why it works: Turns rumination into healthy reflection by training the mind to reframe distorted thoughts.

4. Mindful Micro-Meditation (2 minutes)

Close your eyes, sit still, and notice your breath.

Each time a thought arises, silently say: "thinking."

Gently return to the breath.

Do this for just 2 minutes to reset your brain.

Why it works: Strengthens mindfulness circuits that quiet the DMN and increase clarity.

These tools are designed to be quick, practical, and doable in daily life. Each one is like polishing the mirror of the mind for just a moment, keeping the reflection clear instead of distorted.

Key takeaway

The brain is wired for reflection, but it needs training to serve us rather than trap us. With metacognition, balance in the Default Mode Network, and consistent reflective practices, we can live with clarity, emotional balance, and freedom.

Chapter 10 – Reflection and Well-Being: Why Looking Inward Makes Us Happier

The Mood Mirror

Imagine stepping into your bathroom each morning and facing not just a reflection of your face, but also a reflection of your mind. Instead of showing only hair out of place or tired eyes, the mirror glows softly with shifting colors that represent your current state of mind. A deep red glow signals stress; a hazy gray suggests mental fog; a calm blue shows balance. The mirror doesn't scold, diagnose, or judge; it simply reflects what's happening inside, like a modern-day oracle of awareness.

At first, people treat the mood mirror like a novelty. They laugh when it reveals how tense they are before work or how restless they feel after a night of poor sleep. But over time, the mirror becomes more of a tool for reflection. Instead of rushing past emotions, people pause, breathe, and ask themselves: Why am I feeling this way? What small step could I take to shift?

A teacher sees a sharp red glow one morning before school. Instead of snapping at her students later that day, she uses the mirror's reflection as a reminder to take five deep breaths before class begins. A parent notices their mirror clouded in gray and decides to step outside for a walk in the park instead of numbing out on their phone. A teenager, usually overwhelmed by social media pressures, learns to notice when the mirror shows anxiety rising and chooses to put the phone down before spiraling.

The power of the mood mirror isn't in the technology itself; it's in what it symbolizes. By turning inner states into visible reflections, it reminds people that the mind is not fixed. Stress, sadness, or restlessness are not permanent conditions;

they are like colors or weather patterns, shifting with awareness.

As mood mirrors spread, workplaces install them in lobbies and break rooms. Instead of hiding stress, employees begin talking openly about it. Leaders learn that creating a balanced environment is not just about productivity but about mental clarity. Over time, the mirrors help normalize something ancient traditions always taught: awareness of the mind is the starting point for peace.

In this future, reflection is no longer an afterthought or a luxury; it is woven into daily life. Just as people once used mirrors to polish their outward appearance, now they use mood mirrors to polish their inner state. The result is not just calmer individuals but a society that sees well-being as a shared responsibility. The mood mirror story may belong to the future, but its message is already alive in the present. When we stop long enough to notice our inner state, we gain the power to change it. Science is steadily showing what ancient wisdom has long taught: reflection is not only about knowing ourselves, but also about healing ourselves. It reduces stress, eases anxiety, lifts depression, and builds resilience. Furthermore, reflection enables us to live with a sense of meaning, clarity, and joy.

Mindfulness and Mental Health

In the 1970s, Jon Kabat-Zinn developed what's now known as Mindfulness-Based Stress Reduction (MBSR), blending meditation with modern psychology. Dozens of studies have since shown that mindfulness lowers cortisol levels, reduces physical symptoms of stress, and helps people feel calmer even in difficult situations. Reflection doesn't remove challenges, but it changes the way we meet them.

Instead of reacting automatically, people who practice awareness develop space between the trigger and their response. That space, even if it's just a breath, allows a calmer, wiser choice. Over time, this simple shift transforms how the body and mind handle stress.

Reflection also plays a critical role in treating depression and anxiety. Research shows that mindfulness training can reduce relapse rates in depression by nearly half. For anxiety, reflection shifts focus from spiraling "what ifs" to the grounding reality of the present moment. When someone reflects on their own thoughts with awareness, they begin to notice: This is just worry. This is not reality. That awareness weakens the grip of fear. In this way, reflection doesn't erase negative feelings, but it prevents them from becoming permanent prisons.

Resilience isn't about never struggling; it's about bouncing back. Reflection builds resilience by helping people reframe challenges as temporary and workable. Studies show that those who engage in regular mindfulness or reflective journaling recover more quickly from setbacks, both emotionally and physically. It's not magic, it's training the brain to see storms as passing weather instead of permanent forecasts.

Narrative Psychology

Psychologists have found that humans are natural storytellers. We don't just live experiences; we weave them into narratives. Reflection helps us look at these stories and ask: What do they mean? A hardship can be remembered as a wound or reframed as a turning point. When we pause to reflect on our personal stories, we see patterns. We see how certain choices shaped us, and how we might choose differently moving forward. Our stories become mirrors not just of what happened, but of who we are becoming.

Narrative psychology shows that when people reframe their struggles as opportunities for growth, their well-being improves. For example, someone who once saw a job loss as failure may, upon reflection, reinterpret it as the doorway to discovering a more fulfilling career. Reflection doesn't change the event itself; it changes the meaning we attach to it. And meaning is what makes life livable.

Therapists often use reflective storytelling to help clients reclaim a sense of self. By writing, speaking, or even drawing their stories, people begin to integrate past wounds into a larger whole. Instead of being defined by trauma, they see themselves as survivors, learners, or even guides for others. Reflection turns pain into purpose.

Healthy vs. Unhealthy Reflection

Of course, not all reflection is helpful. When we replay mistakes endlessly, criticize ourselves without mercy, or fixate on "what ifs," reflection becomes rumination. Rumination deepens stress and depression, making us feel stuck rather than free.

Healthy reflection, by contrast, looks at the same situation and asks, "What can I learn?" How can I grow? Instead of endless looping, it moves forward. Constructive reflection is solution-focused, compassionate, and curious. It doesn't avoid hard truths, but it doesn't wallow in them either.

Studies suggest several strategies for breaking out of rumination:

Labeling thoughts: simply saying "this is rumination" helps the brain step back.

Shifting perspective: imagining how a kind friend would view the situation.

Changing context: moving the body, stepping outside, or writing down worries instead of recycling them in the mind.

With practice, these small shifts can turn unhealthy spirals into opportunities for growth.

Practices for Everyday Well-Being

Writing down thoughts isn't just about venting; it's about seeing clearly. Reflection on paper helps us spot patterns, question assumptions, and create distance from unhelpful stories. Even five minutes of journaling can turn mental clutter into insight.

The practice of gratitude is one of the most studied tools in positive psychology. Writing down three good things each day can measurably increase happiness and reduce depressive symptoms for months. Gratitude doesn't deny problems; it balances them by shining light on what is good and present right now.

Happiness doesn't just come from big life changes. It comes from the small moments when we pause, breathe, and notice life as it is. Whether it's a sip of tea, the sound of birds outside, or the smile of a loved one, reflection turns ordinary moments into sources of joy.

Reflection Tools

1. The Three Good Things Practice
Each evening, write down three things that went well that day.

They don't have to be big; even a warm cup of coffee or a smile from a stranger counts.

After writing, pause for a moment and feel the gratitude in your body.

Why it works: This simple practice has been shown in studies to increase happiness and reduce depressive symptoms for months.

2. Story Reframing Journal

Pick one recent struggle or setback.

Write down the story as you currently tell it.

Then, rewrite it from the perspective of growth: What did I learn? How might this make me stronger? How could this help someone else someday?

Why it works: Turns rumination into constructive reflection by shifting focus from pain to purpose.

3. The Rumination Interrupt

When you notice yourself replaying the same worry or mistake, pause.

Silently say: "This is rumination."

Ask yourself: "What's one small action I can take now?"

If no action is needed, redirect by standing, moving, or focusing on your breath.

Why it works: Breaks looping thought patterns and trains the brain to shift into healthier reflection.

4. Mindful Pause Moments

Choose three natural pauses in your day (before meals, while waiting in line, or when switching tasks).

Stop for one deep breath. Notice your surroundings, your body, and one thing you're grateful for.

Then continue with your day.

Why it works: Builds micro-moments of happiness into daily life and strengthens resilience over time.

These tools are designed to make reflection practical and immediate. Small, consistent habits can rewire the brain to shift from stress and rumination into calm, meaning, and well-being.

Key Takeaway

Looking inward isn't about escaping the world; it's about meeting it with more strength, wisdom, and presence. When reflection becomes part of daily life, happiness is no longer a distant goal. It's already here, waiting in the present moment.

Chapter 11 – Reflection and Relationships: Mirrors of Each Other

The Empathy Glasses

The argument started like so many others, voices raised, words sharper than they were meant to be, both people convinced the other "just didn't get it." In the past, this moment might have spiraled into anger, silence, or resentment. But now, they paused and reached for a tool sitting quietly on the counter: a pair of Empathy Glasses.

Lightweight and sleek, the glasses looked like ordinary eyewear. But the moment they slipped them on, something shifted. Through subtle overlays, the glasses highlighted the small emotional cues most people miss: the tremor in a voice, the quickening of a pulse, the tightness in a jaw. A soft glow appeared in the field of vision when the other person's stress levels rose; a gentle ripple of color pulsed when sadness, fear, or tenderness was detected beneath the surface.

Instead of just seeing anger, they saw what lived under it: fear of not being heard, a longing to be understood, a hurt carried from earlier in the day. The glasses didn't fix the problem or tell anyone what to do. They simply reflected reality more clearly, peeling back the distorted layer of assumptions and projections.

For the first time, instead of reacting to anger with anger, they paused. One said softly, "You're scared I don't value your opinion, aren't you?" The other's eyes widened, softened because yes, that was it, and finally someone saw.

In that moment, the fight dissolved not because the conflict was solved but because the connection returned.

Soon, these Empathy Glasses began to appear in workplaces, classrooms, and even community forums where debates once exploded into division. Strangers who might have shouted past each other paused when the glasses revealed fear instead of hatred, grief instead of aggression. Slowly, a cultural shift took place.

It wasn't magic. People still disagreed, still felt anger, still carried old wounds. But the glasses reminded them of something reflection has always offered: that what we see on the surface is rarely the whole truth. Behind every harsh word is a feeling; behind every outburst, a story.

And in seeing more clearly, people remembered something even more powerful: that connection begins with understanding. The Empathy Glasses were just technology; the real change came when people began to reflect more deeply on what others were truly showing them.

The Empathy Glasses may be a futuristic idea, but the truth they represent has always been part of us. Human beings are walking mirrors. We don't just live in isolation; we echo one another, consciously and unconsciously. From the smallest smile exchanged with a stranger to the deepest conflict in a family, relationships shape who we are and how we feel. And reflection is the key that determines whether those connections heal us or hurt us.

The Neuroscience of Connection

In the 1990s, neuroscientists discovered mirror neurons, brain cells that fire both when we perform an action and when we observe someone else performing it. If you watch someone smile, the same neural circuits light up in your brain as if you were smiling yourself. If you see someone cry, your brain mirrors their sorrow.

This is the biology of empathy. It means our nervous systems are not separate islands but tuned instruments, resonating with one another. We feel what others feel sometimes before words are even spoken. This mirroring doesn't stop at gestures; it extends to emotions as well. Research on emotional contagion shows that moods spread like invisible currents. A stressed boss creates a stressed office. A calm teacher creates a calm classroom. A joyful laugh ripples through a crowd.

Reflection here is not abstract; it's embodied. When we are unaware, we unconsciously absorb the moods of those around us. When we are aware, we can choose whether to pass them along or transform them.

Because emotions spread so easily, the first responsibility is to notice what we ourselves bring into the mirror. Self-awareness interrupts the chain reaction. Instead of snapping back with anger when someone is upset, reflection allows us to ask, 'What am I feeling right now?' What state of mind am I reflecting into this space?

Relationships thrive not because conflict disappears but because reflection turns reaction into choice.

How Reflection Improves Communication

One of the simplest yet most powerful skills in relationships is reflective listening: pausing long enough to repeat back what we heard. "So what I'm hearing is that you felt ignored when I didn't respond." This pause interrupts reactivity and creates space for empathy. Neuroscience shows that when people feel heard, their nervous systems calm down. Reflection in communication isn't just polite; it's a biological regulation.

Much of what we accuse others of saying or feeling is actually our own projection. Reflection helps us see this: Am I angry at them, or am I angry at the story I'm telling myself about them? When we own our projections, we clean the mirror and stop blaming others for the smudges on our own glass.

All relationships experience rupture. What matters is whether we repair. Reflection turns repair into an active practice: noticing our part in the conflict, acknowledging it, and choosing new words or actions to address it. When both sides reflect honestly, conflicts become stepping stones to deeper trust rather than cracks that widen over time.

Forgiveness and Connection

Forgiveness is not pretending the harm never happened. It begins with reflection: What story am I replaying? How long do I want to carry this pain? Self-awareness helps us distinguish between protecting ourselves from real harm and clinging to resentment that corrodes us from within. Studies show that forgiveness lowers blood pressure, reduces stress, and improves mental health. Forgiveness is not only spiritual wisdom, but also a form of biological medicine. Letting go clears the inner mirror, allowing us to see others as more than their mistakes.

Compassion as the Natural Result of Clarity

When the mirror of the mind is clear, compassion arises naturally. We see that everyone struggles, everyone carries wounds, and everyone makes mistakes. Compassion does not excuse harmful actions, but it allows us to respond with humanity rather than endless cycles of retaliation.

The Ripple Effect on Communities

A parent who pauses to reflect before reacting to a child's misbehavior models emotional regulation. A partner who listens instead of snapping creates a safe space for intimacy and connection. These small acts ripple outward, shaping the atmosphere of the home.

Healthy families become the building blocks of healthy communities. A neighborhood where people know how to listen, forgive, and reflect together is a place where cooperation is natural and conflict doesn't spiral out of control.

At the broadest level, reflection in relationships scales into reflection in societies. Suppose individuals can see others not as enemies but as mirrors. In that case, collective decisions are guided less by fear and more by empathy. A society that values reflection is one that builds bridges instead of walls, and peace instead of endless division.

Reflection Tools

1. Reflective Listening Drill

Pair with a friend, partner, or family member.

One person speaks for 2 minutes about something meaningful.

The listener may not interrupt — only reflect back what they heard. ("So you felt ___ when ___ happened.")

Switch roles.

Why it works: Builds empathy by training you to hear feelings, not just words.

2. Projection CheckIn

When upset with someone, pause and ask:
What story am I telling about them?
Is this story fact, or is it my interpretation?
What part of me might be reflected in this judgment?
Why it works: Reduces blame by distinguishing between the other person's behavior and your own projections.

3. Forgiveness Reflection Journal
Write down someone you feel resentment toward.
Ask yourself: What am I holding onto, and what is it costing me?
Write a short statement of release: "I don't condone what happened, but I choose not to carry this pain any longer."
Repeat as often as needed.
Why it works: Shifts energy from clinging to resentment toward clarity, healing, and freedom.

4. Compassion Pause
When emotions rise in a conflict, pause and silently repeat: "Just like me, this

person wants to be safe, understood, and loved."

Take one deep breath before responding.

Why it works: Activates compassion in the moment, softening reactivity and opening the door to connection.

These practices are designed to strengthen relationships at every level, from personal partnerships to communities. Reflection in connection doesn't just heal individuals; it builds families and societies rooted in empathy.

Key Takeaway

We are all mirrors of one another. When we reflect clearly, we bring clarity into our relationships. When we cloud the mirror with fear, blame, or projection, we multiply distortion. However, the choice is always available: to pause, reflect, and meet others with empathy. In doing so, we not only heal our closest bonds but also help shape a more compassionate world.

Chapter 12 – Collective Reflection: When Groups Look in the Mirror Together

The Collective Heartbeat

The room was filled with strangers, people who had come together for a dialogue about a divisive community issue. At first, the tension was palpable: folded arms, guarded faces, cautious glances. Each person was carrying their own story, their own hurt, their own perspective; they were ready to defend. Then the facilitator invited everyone to put on small, lightweight wristbands. These bands, linked to a central display, began tracking each person's heart rate and breathing. On the wall, a soft ripple of lights appeared, each pulse a flicker of color. At first, the pattern was chaotic: quick, sharp flashes from anxious participants, followed by slower waves from those who were more relaxed.

"Now," the facilitator said gently, "let's take a breath together."

One inhale. One exhale. Slowly, the lights on the wall began to sync. Dozens of separate pulses softened into a single rhythm, a wave rising and falling as one. The room grew quiet. The guarded looks shifted into curiosity. For a moment, everyone could see and feel what was usually invisible: that beneath differences, every heart beat with the same steady rhythm.

As the conversation began, something was different. People still disagreed, but the anger softened. They listened longer and responded with less hostility. The visualization of the collective heartbeat reminded them: we are connected, even when we forget.

In time, these "heartbeat circles" spread to schools, workplaces, and even political forums. Communities discovered that when people first felt their physical connection, it became easier to engage in reflective dialogue. Arguments became less about "me versus you" and more about "us, trying to understand together." The technology wasn't magic; it didn't erase conflict. But it created a shared mirror, one that reflected not words or beliefs, but life itself. And in that mirror, people saw what they too often missed: that every pulse, no matter whose, carries the same longing for safety, dignity, and belonging.

The vision of a room full of strangers syncing their heartbeats may seem futuristic. Still, it illustrates something deeply human: when people reflect together, barriers begin to dissolve. Shared reflection doesn't erase conflict, but it changes the way conflict is carried. Instead of seeing only separation, we glimpse connection. And when communities reflect as one, the possibilities for empathy, healing, and collaboration grow exponentially.

The Psychology of Shared Reflection

Psychologists have long studied the power of perspective-taking, the act of imagining the world through another person's eyes. Research shows it can reduce prejudice, soften stereotypes, and increase empathy across cultural and political divides. Shared reflection creates space for this process, encouraging individuals to pause, step outside their assumptions, and recognize the humanity in others.

When groups intentionally practice perspective-taking, the "us versus them" wall thins. Instead of seeing an opponent, people begin to see another human with hopes, fears, and struggles not so different from their own.

Group conversations often escalate because of defensiveness. When someone feels attacked, they push back harder. But reflection shifts these dynamics. When participants are guided to pause, listen, and reflect on what they've heard before responding, conversations become less about winning and more about understanding.

Research on group facilitation shows that reflection slows the tempo of dialogue, lowers aggression, and increases cooperation. Even in heated debates, reflection acts like a cooling agent, helping groups navigate complexity without collapsing into hostility.

Empathy is not just an individual trait; it can function as a collective skill. Studies in social psychology show that when groups reflect together, levels of shared empathy rise. This is sometimes referred to as "collective empathy," and it helps explain why communities that practice reflection together tend to report higher levels of trust and cooperation.

In other words, empathy can be scaled. It doesn't have to stop with individuals; it can become part of a group's culture.

Practices That Bring Communities Together

One of the most effective modern practices for collective reflection is the dialogue circle. Used in schools, workplaces, and justice systems, these circles bring together those harmed, those responsible, and community members to share openly. Instead of punishment, the focus is reflection: What happened? How were people affected? What needs to be done to make it right?

Studies show restorative justice circles reduce repeat offenses and increase empathy for all involved. Reflection replaces alienation with connection, even in the face of harm.

Mindfulness is often practiced alone, but group mindfulness sessions are gaining traction. When a room of people breathes together, sits in silence together, or shares moments of stillness, a collective calm emerges. This practice has been adapted for classrooms, prisons, and even corporate boardrooms.

The impact is powerful: people feel less alone in their stress and more grounded in the shared experience of simply being human.

Every community carries wounds, whether from conflict, inequality, or trauma. Story-sharing provides a path to heal. Programs like story circles invite individuals to share personal experiences while others listen reflectively, without interruption.

Hearing someone's story firsthand often breaks stereotypes faster than data or debate ever could. Stories turn statistics into people, and people into neighbors. Reflection through storytelling rebuilds trust where it has been fractured.

Scaling Reflection from Individuals to Societies

Healthy communities require more than voting; they require citizens who can reflect on their own biases, assumptions, and responsibilities. Reflection fosters humility, making space for dialogue instead of polarization. When citizens reflect on their part in the larger whole, societies shift from "my side must win" to "how do we all move forward together?"

Case Studies of Collective Reflection in Movements and Communities

Throughout history, movements for change have used collective reflection as a foundation. Civil rights leaders often gathered for prayer and reflection before marches. Truth and reconciliation commissions, such as those in South Africa, invited entire nations to reflect on past harms as a way of healing. These examples show that when societies reflect honestly, even deep wounds can begin to mend.

Reflection is more than an individual wellness tool; it's a cultural necessity. Societies that normalize collective reflection are more adaptable, cooperative, and compassionate. Instead of being driven solely by fear and reaction, they act from clarity and connection. In a rapidly changing world, this capacity may be one of the most valuable tools for survival that humanity possesses.

The Future of Collective Wisdom

Technology could become a bridge for collective reflection. Virtual "reflection forums" may one day allow people from across the globe to pause, share stories, and listen in real time. Just as social media connected the world with speed, reflective platforms could connect the world with empathy.

Even as technology advances, many of the most powerful practices come from the past. Hoʻoponopono in Hawaii, vision quests in Indigenous traditions, storytelling circles in Africa, all remind us that reflection has always been communal. Reviving these practices in modern forms could provide the balance our hyperconnected but lonely world desperately needs.

Imagine societies where schools teach collective reflection as a civic skill, workplaces embed it into culture, and communities gather regularly to reflect together. In such a world, diversity would not be seen as a threat but as an opportunity to see more clearly through many mirrors. Reflection would no longer be a rare skill; it would be the heartbeat of society itself.

Reflection Tools

1. Dialogue Circle at Home or Work

Gather 3–6 people (family, friends, coworkers).

Sit in a circle and choose one prompt (e.g., "What's one challenge I faced this week and what did I learn?").

Each person speaks without interruption; others only listen.

After everyone shares, reflect back on one thing you learned from the group.

Why it works: Builds empathy, reduces defensiveness, and strengthens trust.

2. Shared Mindful Breathing (5 Minutes)

In a group, sit together in silence.

Breathe slowly in unison (a facilitator can guide with counts).

At the end, each person shares one word describing how they feel.

Why it works: Calms group dynamics and creates a sense of shared presence.

3. Story Exchange Practice

Pair up in a group setting.

Each person tells a short personal story about a meaningful life event (5 minutes each).

Switch, then reflect back on what stood out in the other's story.

Rotate partners if in a larger group.

Why it works: Story-sharing humanizes others, breaks stereotypes, and builds community.

4. Community Gratitude Wall

In schools, workplaces, or neighborhoods, set up a board where people can anonymously write what they appreciate about others.

Encourage adding reflections weekly. Read them aloud at the end of the month in a group gathering.

Why it works: Shifts focus from problems to appreciation, strengthening collective bonds.

These practices take reflection beyond the individual and into the group, making it a living, shared experience. Done consistently, they can transform not just relationships but entire communities.

Key Takeaway

When individuals reflect, they heal themselves. When groups reflect, they heal relationships. When societies reflect, they heal history and open futures. Collective reflection is not easy, but it is possible. And if adopted, it could be one of humanity's greatest tools for building a more compassionate and connected world.

Chapter 13 – The Future of Reflection Research: Technology Meets Consciousness

The Shared Brainwave Experiment

The auditorium was quiet as twenty strangers sat in a circle, each wearing a lightweight headset that looked no different from a pair of over-ear headphones. Electrodes, hidden in the fabric, were reading their brain activity. A large screen at the front of the room displayed shifting lines of light, jagged and uneven, each representing the brainwaves of a different participant.

At first, the lines moved chaotically, no two alike. Some pulsed quickly, reflecting restless thoughts; others dipped and rose erratically, signs of stress or distraction. The group had never met before. Some were skeptical, some nervous, some curious.

Then the facilitator invited everyone to close their eyes. "Take one breath together," she said softly, "and another. Let the mind settle, like dust in water."

As they breathed, something remarkable began to happen. The jagged lines on the screen slowly softened. Waves that had been separate began to fall into rhythm. Within minutes, clusters of lines rose and fell almost as one. By the ten-minute mark, the majority of the group's brainwaves had synchronized into a shared, steady pattern.

The participants didn't know this at the moment. What they felt was simpler: a strange calm, a sense of being understood without speaking, an ease that hadn't been there when they walked in. Some described it later as "being part of a single heartbeat." Others said it felt like the invisible walls between them had thinned.

This was the Shared Brainwave Experiment, one of the first large-scale attempts to measure collective consciousness. Scientists wanted to know: What happens when people reflect together, not just with words, but with their minds? Could group reflection produce measurable changes in the brain? The early findings suggested yes. Synchrony, the alignment of brain activity across individuals, appeared most strongly when groups practiced mindfulness, storytelling, or compassionate reflection. The more aligned the group's brainwaves, the more participants reported feelings of empathy and connection afterward.

The implications were staggering. If reflection could shift not only individual brains but groups of brains, then perhaps ancient wisdom traditions were pointing to a scientific truth: we are not just separate selves but part of a larger field of awareness.

Imagine classrooms where students practice brief mindfulness sessions and their brainwaves begin to harmonize, creating a more focused and collaborative learning environment. Imagine peace talks where negotiators breathe together before discussions, synchronizing minds before debating terms. Imagine communities measuring their "collective calm" the way we now measure air quality, not as a gimmick, but as a sign of social health.

The Shared Brainwave Experiment didn't provide all the answers, but it opened a door. It suggested that reflection is not only personal medicine but a communal force, capable of aligning not just hearts but minds. And in that alignment, perhaps, lies the future of connection.

The study of reflection is no longer confined to monasteries, philosophy halls, or psychology labs. Today, it extends into the glowing screens we hold, the sensors we wear, and the artificial intelligence systems that are beginning to act as mirrors for our inner worlds. Ancient wisdom has long described the mind as a reflective surface, but now technology is testing that metaphor in real time, opening doors that humanity has never walked through before. The question is not just what these tools can show us, but how we choose to use them and whether they will deepen our humanity or distract us from it.

AI as a Reflective Companion

Journaling has always been a practice of reflection, but with the advent of AI, the process takes on a new shape. Instead of only recording thoughts, AI systems can analyze them, spotting patterns in language that a person may overlook. For example, someone who repeatedly writes about failure may not realize how often self-criticism dominates their entries. An AI journaling companion can highlight those blind spots, suggesting reframes or drawing attention to hidden strengths. In this way, technology becomes not just a notebook, but a mirror that reflects the patterns of the mind back to the writer. The rise of AI-driven "mindfulness coaches" points toward a future where reflection is not practiced in solitude alone but supported by interactive digital companions. These coaches can guide breathing exercises, prompt moments of self-inquiry, and even respond with compassion-trained language when users express distress. Imagine a system that gently reminds you when your attention has drifted to rumination, nudging you back toward clarity. The potential here is immense, not as a replacement for human teachers, but as an accessible entry point to daily awareness.

This new landscape also raises serious questions. When AI learns to act as a mirror, who holds the data being reflected? Should a private moment of confession to an algorithm be treated as securely as a confession to a human counselor? There are risks of dependence, surveillance, or the commercialization of inner life. The very intimacy that makes AI reflection tools powerful also makes them vulnerable to misuse. Technology may sharpen the mirror, but without ethics, it could easily distort it.

Wearables and Neurofeedback

Wearables, ranging from smartwatches to brainwave headbands, already measure heart rate, sleep patterns, and stress levels. They are beginning to expand into deeper territory: tracking subtle emotional states. A sudden spike in heart rate variability or a shift in brainwave activity can act as a signal that the inner mirror is clouded. These devices allow people to see stress not just as a vague feeling, but as a measurable condition of the body-mind system.

The real power of these tools lies not only in detection but in real-time intervention. Neurofeedback devices can alert someone when they are slipping into an anxious loop, guiding them back with sound cues, breathing prompts, or calming visuals. In effect, the mirror becomes dynamic, not just reflecting a cloudy state, but also helping to polish it as it appears. This transforms awareness from a delayed reflection into an immediate practice.

The implications for mental health are profound. While therapy and counseling are limited by time and access, wearable devices could act as everyday allies, offering gentle corrections before distress spirals into crisis. Though not a replacement for professional care, they can democratize access to reflective tools, making daily well-being more attainable for millions.

Collective Neuroscience

Reflection is not only individual, it is also collective. Recent studies in collective neuroscience show that when people meditate, sing, or even listen to a story together, their brainwaves begin to synchronize. The mirror of the mind extends beyond one skull, forming a shared rhythm in groups. This suggests that collective practices of reflection, from dialogue circles to communal rituals, are rooted not just in culture but also in biology.

In groundbreaking experiments, researchers have measured how one person's state of mind directly influences another's through brain-to-brain synchrony. When someone listens with genuine empathy, the listener's brainwaves often begin to mirror the speaker's. This scientific validation of empathy reveals that reflection is not an isolated process but one that ripples outward, literally connecting minds in shared states.

If these findings are confirmed, we may one day understand how collective consciousness operates, whether in small groups or at societal levels. Could synchronized states of compassion shift cultural conflicts? Could collective practices of reflection be measured, cultivated, and amplified in ways that strengthen global unity? Science may not yet have the tools to answer fully, but the doorway has opened.

Risks and Possibilities

The future of reflection research holds enormous promise. Technology may offer people new ways to see themselves clearly, break free from cycles of rumination, and cultivate empathy at scales once unimaginable. AI companions, wearables, and collective neuroscience could become as common as fitness apps, helping humanity develop mental fitness and emotional clarity. But with this promise comes danger. If people outsource reflection entirely to devices, they risk losing the very inner capacity those tools were meant to support. If corporations or governments misuse reflective data, privacy and freedom could be compromised.

Technology can sharpen the mirror, but it can also warp it.

The challenge, then, is balance. Ancient wisdom traditions remind us that the mirror of the mind requires stillness, honesty, and humility, qualities no device can fully provide. Modern innovation offers reach, accessibility, and measurement that traditions could never achieve alone. The future of reflection research will not be found in choosing one over the other, but in weaving them together, so that technology amplifies wisdom without replacing it.

Reflection Tools

1. AI-Assisted Journaling

Use AI journaling apps to analyze your daily entries.

Ask prompts like: "What themes am I repeating?" or "What blind spots do you notice in my reflections this week?"

Keep control of your privacy: export, delete, or encrypt data when possible.

2. Mindfulness & Virtual Coaches

Try guided mindfulness apps. Some already use adaptive coaching that adjusts prompts based on your usage.

Set reminders for "micro-reflections" throughout the day (1-minute pause to notice breath before meetings).

Treat them as supportive companions, not replacements for human connection.

3. Wearables for Emotional Awareness

Smartwatches can track heart rate variability (HRV), a marker of stress and resilience.

Brain-sensing headbands (Muse, Emotiv) provide neurofeedback for meditation, showing when your mind drifts.

Use alerts as "mirror signals" not judgments, but reminders to pause, breathe, and reset.

4. Neurofeedback Practices

Try apps and devices that use light/sound feedback when your brain activity shifts into stress patterns.

Example practice: notice the alert, take three deep breaths, and label the thought "just a reflection".

Over time, you train your brain to return to balance faster.

5. Group Reflection Experiments

Use group journaling platforms (like 750words or collaborative Notion pages) for collective reflection.

Join virtual meditation or storytelling circles where participants share experiences, noticing the resonance that comes from synchrony.

In communities, experiment with "reflection rounds": 5 minutes each, where one speaks and others only listen.

6. Digital Hygiene for Reflection

Decide which reflections will stay analog (private journal, meditation) and which you will allow tech to assist with.

Protect your "mirror space": turn off notifications during reflection practices.

Regularly ask: "Am I using this tool to deepen awareness or to distract myself?"

Key Takeaway

Technology can sharpen the mirror, but it's up to you to decide when to pick it up and when to put it down.

Chapter 14 – Education for Reflection: Teaching the Next Generation

The Story Mirror

Imagine walking into a quiet room after a long, stressful day. You sit down in front of a simple device, a screen framed like a mirror. Instead of showing only your reflection, it invites you to speak. You begin telling your story: how your boss dismissed your idea in the meeting, how the bills keep piling up, how you feel stuck, small, and unseen.

At first, it feels like venting into a diary. But then something happens. The "mirror" listens not just to your words, but to your tone, and your patterns, the threads running underneath the surface. After you finish, it reflects your story back to you, not as a cold analysis, but as a new perspective.

The harshness you carried in your voice is softened into compassion: "You showed courage by speaking up today, even though your idea wasn't heard. That doesn't mean it wasn't valuable. Your persistence is a strength, not a weakness." The feelings of failure and frustration are reframed into possibility: "You're not stuck. What you're facing is a chapter of challenge, and every story has those. The fact that you're reflecting now means you're already growing."

You sit back, startled. For the first time all week, you don't feel crushed by your thoughts. You feel lighter. Not because the challenges disappeared, but because you see them differently. The "mirror" didn't magically fix your problems; it helped you see the truth beneath the distortion.

Now imagine this practice scaled beyond one room. Workers step into reflection pods before leaving the office, turning stress into clarity. Couples use story mirrors at home to process conflicts before words harden into walls. Leaders use them before making major decisions, checking not just data but their own mindset. Communities gather around shared story circles where AI tools reflect collective fears, hopes, and strengths, giving groups a chance to step out of blame and into understanding.

The future of reflection might not look like silence on a mountain or ink on paper. It might also look like technology becoming a compassionate witness, a mirror that listens deeply and gently hands back a clearer story.

Because often, the biggest shift doesn't come from changing our circumstances. It comes from changing the story we tell ourselves about them. And if that story becomes kinder, truer, and more empowering, then the world outside changes too.

The classrooms of the future will not only teach equations and grammar, but also instill presence, compassion, and the art of reflection. For centuries, education has focused on the transfer of knowledge; however, we are now beginning to realize that knowledge alone is insufficient. Without wisdom, knowledge can be misused. Without self-awareness, even the brightest minds can stumble. Without reflection, learning remains shallow facts memorized but never integrated into life. Reflection, when taught as a skill, offers the possibility of nurturing whole human beings, not just test-takers. It prepares people not only for careers, but for life itself.

Reflection in the Classroom

Across the globe, schools are beginning to experiment with reflection-based practices. These are not abstract philosophies but practical tools woven into the rhythm of the school day.

One example is mindfulness. In some classrooms in the United States, the United Kingdom, and India, children start the day with a short breathing exercise, two or three minutes of quiet, where the focus shifts from rushing thoughts to the sensation of the breath. For many, this simple pause reduces anxiety, steadies attention, and gives a sense of calm before lessons begin.

Another growing movement is social-emotional learning (SEL), now taught in schools in over 30 states in the U.S. and spreading internationally. SEL explicitly teaches students how to recognize their emotions, empathize with others, resolve conflicts, and reflect before reacting. Instead of treating emotions as distractions, SEL treats them as signals that can be understood and navigated through awareness.

Early evidence is promising. Schools that incorporate mindfulness or SEL often report fewer disciplinary issues, reduced bullying, better classroom focus, and improved overall well-being. Teachers describe students as more capable of listening, less reactive in conflicts, and more willing to help each other. The lesson is clear: when reflection is part of learning, children don't just do better in school, they do better in life.

The Science of Reflective Learning

The science behind these results is becoming clearer. Mindfulness has been shown to improve focus and attention by strengthening the prefrontal cortex, the brain's "executive center" responsible for planning, regulating emotions, and decision-making. Just a few minutes of daily practice can help students tune out distractions and sustain concentration. Reflection also builds empathy. Studies show that children who engage in mindfulness and narrative reflection exercises are more likely to act with kindness and cooperation. This is partly due to changes in the brain's mirror neuron system, which helps us "feel with" others. By teaching reflection, we are quite literally strengthening the brain's capacity for compassion.

Perhaps most importantly, reflection nurtures resilience. Children who learn to pause and reframe their experiences, rather than being overwhelmed by them, are more adaptable to stress. When setbacks come, they are more likely to see challenges as learning opportunities rather than threats. This adaptive capacity may be one of the most important skills of the 21st century, where rapid change is the norm.

Imagining Reflection as a Core Skill

Imagine a world where reflection is taught alongside math, reading, and science. Where every student graduates not only with academic knowledge but with the ability to pause, breathe, and respond with clarity rather than reactivity.

Reflection as a core skill would transform education. Students would learn how to listen to themselves as much as to their teachers. Classrooms would become spaces where curiosity is paired with compassion, and where achievement is measured not only by grades but by growth in empathy, resilience, and wisdom.

In such schools, children would grow up understanding that they are more than their mistakes, that others are not their enemies, and that challenges are invitations to grow. Reflection would not be an extra activity for a few minutes on Fridays, but a daily practice embedded in how lessons are taught and how problems are solved.

This vision is not limited to children. Adults, too, can benefit from schools and workplaces designed with reflection at their core. In communities where reflection is valued, meetings begin with silence, decisions are weighed with compassion, and disagreements are resolved with understanding rather than aggression.

The ultimate goal is inner and outer growth, a generation capable of navigating both personal life and global crises with clarity and insight.

Preparing the Next Generation of Leaders

Why does this matter so much? Because the challenges ahead are not small. Climate change, technological disruption, global inequality, and cultural division demand leaders who can see clearly, think deeply, and act with compassion. Without reflective skills, leaders may default to fear, reaction, and short-term thinking. With reflection, they can approach problems with patience, wisdom, and the ability to hold multiple perspectives at once.

Cultivating reflective decision-makers means preparing a generation of leaders who are not only intelligent but wise. Leaders who can pause before acting. Leaders who can recognize when ego is clouding their judgment. Leaders who can see their opponents not as enemies but as

fellow human beings with their own
reflections of reality.

In such a world, societies would no longer
value knowledge alone, but wisdom
equally. Wisdom would not be seen as
something mystical or rare, but as a
natural outcome of teaching reflection
from a young age.

Reflection Tools

1. One-Minute Pause (Classroom or
Workplace)

At the start of a lesson, meeting, or day,
set aside 60 seconds for silence.

Focus only on the breath, the feeling of
sitting, or sounds in the environment.

This tiny pause creates calm, improves
focus, and trains the brain to switch from
reactivity to awareness.

2. Reflection Journals

Encourage students or employees to keep
a simple notebook with two daily
prompts:

"What did I learn or notice today?"

"How did I grow or connect today?"

Writing even two or three sentences builds the habit of reflection and shows that growth is not just about achievements, but about awareness.

3. Story Circles

Groups sit together and share a short story from their day or life.

Rules: listen without interruption, no judgment, no advice.

This builds empathy, teaches perspective-taking, and mirrors Indigenous and African traditions of storytelling as collective reflection.

4. Mindful Transitions

Between classes, after lunch, or before a new task, practice a 30-second breathing reset.

Example: "Inhale calm, exhale tension."

This practice reduces stress and keeps the "mirror of the mind" clear throughout the day.

5. Gratitude Wall (or Board)

Dedicate space in the classroom or office for people to write one thing they're grateful for each day.

Seeing gratitude reflected back as a group builds positivity, belonging, and shared perspective.

6. Role-Play Reflection

Practice real-world challenges through role-play, then pause to reflect.

Questions: What did you feel? What assumptions did you notice? How might you respond differently next time?

Helps children and adults practice reflective responses rather than automatic reactions.

7. The Mirror Question

Before making a choice, ask: "If I look in the mirror tonight, will I feel aligned with this decision?"

This simple habit builds self-awareness, integrity, and long-term thinking.

8. Community Reflection Days

Once a month, schools or workplaces set aside time for shared silence, storytelling, or gratitude practice.

Creates a culture where reflection is not a side activity, but part of the collective identity.

Key Takeaway

When reflection becomes part of education, we raise not just more brilliant students, but more compassionate humans. We create classrooms where children learn not only to solve equations but also to solve conflicts. We prepare workplaces where decisions are made not just for profit, but for people. And we build societies where wisdom is not an accident of age, but a cultivated quality of every citizen.

Reflection is not just a tool for personal well-being; it is a foundation for the future of humanity.

Chapter 15 – Reflection and Global Challenges: A Science of Connection

The Global Reflection Summit

The world had grown weary of endless conferences where leaders spoke louder than they listened. Year after year, climate talks stalled, peace negotiations broke down, and global summits ended with vague promises and little change. Something had to shift, not in policy, but in perspective.

So when the first Global Reflection Summit was announced, many were skeptical. "Another gathering of politicians?" critics scoffed. But this summit was different. Instead of opening with speeches, it opened with silence.

On the morning of the first day, hundreds of world leaders, scientists, spiritual teachers, and community representatives from across the globe sat in a vast circular hall. The lights dimmed, and for ten minutes, no one spoke. No cameras clicked. No translators whispered. There was only the sound of collective breathing, a rare pause in a world addicted to noise.

At first, the silence was uncomfortable. Some fidgeted in their seats, minds racing with arguments they wanted to make. But slowly, the room settled. The stillness became palpable, like water smoothing after a storm. And in that stillness, something shifted: leaders began to see each other not as adversaries, but as fellow human beings carrying the weight of their people's hopes.

When the dialogue began, it wasn't business as usual. Each participant wore a small device, known as an "empathy translator," that not only converted language but also conveyed emotional tone. For the first time, a negotiator could hear the fear behind another's sharp words, or the longing hidden in a defensive stance. The device didn't solve disagreements, but it revealed the humanity beneath them.

One leader spoke of rising seas swallowing her island nation. Another shared the despair of young people marching in her capital, demanding change. Instead of countering with arguments, others paused, reflected, and acknowledged the truth of what was said. The summit became less about debating and more about collective problem-solving.

By the end of the week, agreements were signed, not perfect, but real. A new framework for climate responsibility. A shared fund for rebuilding after conflicts. A pledge to teach reflection practices in schools worldwide.

But beyond the agreements, something deeper had changed. Those who attended left with a new perspective: that reflection was not a weakness, but a form of power.

That slowing down could move humanity forward faster than rushing headlong into conflict. That solutions emerge not from domination, but from understanding.

The Global Reflection Summit became an annual event. And over time, its influence spread. Nations began opening negotiations with silence instead of threats. Corporations introduced reflection circles before making high-stakes decisions. Even local communities borrowed the model, gathering to breathe, listen, and speak with clarity before resolving disputes.

It turned out the greatest innovation of the 21st century wasn't a new technology, but a simple rediscovery: when people reflect together, they remember they are connected. And when they act from that connection, the world changes.

When we think about solving global challenges, such as war, climate change, corruption, and inequality, we often think first of strategies, policies, or technologies. But beneath all these tools lies something more fundamental: the human mind. The way we see, interpret, and respond to the world shapes the choices we make as individuals, nations, and a species. Reflection is not just a private act of self-care; it is a collective skill that can decide the future of humanity.

Reflection as a Path to Peace

Conflict often arises not from facts, but from perception. When we fail to reflect, we mistake our assumptions for reality and our fear for truth. Reflection allows us to step outside of immediate emotion and see others not as enemies, but as fellow human beings with their own fears and needs.

Perspective-taking reduces hostility. Neuroscience reveals that when people actively imagine the perspective of another, activity increases in brain regions associated with empathy and understanding. Nations, communities, and even families benefit when people pause long enough to ask: What might this look like through their eyes?

Dialogue and forgiveness have shaped peace-building. From truth and reconciliation commissions in South Africa to local community circles in Rwanda, structured reflection has been employed to heal even the most profound wounds. These processes work not because they erase the past, but because they create space for people to see beyond vengeance into the possibility of coexistence.

Reflection is the foundation of diplomacy. Great negotiations rarely happen in heated outbursts. They happen when leaders step back, breathe, and consider what outcome will create harmony rather than prolong hostility. Reflection is the difference between escalating a conflict and laying the groundwork for peace.

Reflection and Environmental Responsibility

The environmental crisis is not just about carbon, pollution, or waste; it is about perspective. When we see ourselves as separate from nature, exploitation feels justified. But when we reflect, we begin to recognize the truth: the health of the planet is a reflection of the health of humanity.

Nature mirrors our actions. Forests stripped bare, oceans filled with plastic, and rising temperatures are not distant "problems"; they are the outer reflection of inner disconnection. Just as a polluted mind distorts clarity, a polluted world distorts the balance we rely on.

Reflection fosters ecological awareness. Indigenous traditions practiced reflection with nature, fasting, sitting in silence, or offering gratitude. Modern psychology confirms that time spent reflecting outdoors increases compassion, patience, and a sense of connection to life. These states naturally inspire sustainable choices.

Inner clarity leads to sustainability. When the mirror of the mind is clear, impulsive consumption gives way to mindful stewardship. People begin to buy less, waste less, and prioritize long-term balance over short-term gain. A reflective society is also a sustainable society.

Ethical Leadership Through Reflection

Leaders hold immense power. Their decisions ripple across economies, communities, and ecosystems. Without reflection, that power can easily become corrupted by greed, ego, or fear. With reflection, it can become a force for wisdom and care.

Leaders who pause lead wisely. Studies on decision-making have shown that even brief reflective breaks can reduce impulsivity and increase fairness. A leader who takes a breath before responding is more likely to see the bigger picture and avoid destructive escalation.

Reflection safeguards against corruption. The unexamined leader is prone to justify harmful actions, convincing themselves of their righteousness. Reflective leaders, by contrast, continually question their own motives, asking: Am I serving myself, or the greater good?

History shows the power of reflective leadership. Figures such as Nelson Mandela embodied reflection, choosing forgiveness over revenge, reconciliation over domination. These choices were not easy, but they were only possible because of a deep commitment to inner clarity before outer action.

Long-Term Thinking for Future Generations

Perhaps the greatest gift of reflection is that it expands our vision beyond the present. In a distracted, fast-moving world, most decisions are made for immediate comfort or short-term gain. Reflection slows us down long enough to ask: How will this choice affect those who come after me?

Reflection expands perspective. By stepping back, we see not only our own interests but also the impact on others, our families, our communities, and the generations yet to come. Reflection makes us more patient, more cautious, and more compassionate.

Legacy and responsibility become central. A reflective person asks not just, What do I want today? But what am I leaving behind? This shift turns leadership from ambition into stewardship.

Societies thrive when reflection is valued. Imagine a world where leaders are trained not only in economics or law, but also in reflection, where they practice silence before debate, gratitude before policy, and perspective-taking before decisions. That is not weakness; it is wisdom in action.

Reflection Tools

1. The Perspective Pause (Conflict Resolution)

Before responding to conflict (personal, political, or international), pause and ask:

"What might this look like from their side?"

"What need or fear could be driving their behavior?"

This simple pause interrupts defensive escalation and opens space for reconciliation.

2. The Mirror of Nature (Environmental Awareness)

Spend 15 minutes observing a natural space — a tree, river, or even the sky.

Ask: "What is this place reflecting back about how we live?"

Journal insights, especially about consumption, balance, and responsibility.

This practice reconnects inner reflection with environmental care.

3. Legacy Letters (Long-Term Thinking)

Write a short letter to someone living 100 years in the future.

Share what you hope they inherit from your generation — and what choices you believe matter most today.

This shifts reflection from short-term gain to long-term stewardship.

4. Restorative Circles (Community Healing)

Adapted from Indigenous and restorative justice traditions:

Gather participants in a circle.

Each speaks in turn about harm done, needs for healing, and hopes for the future.

No interruptions, no debate — only reflection and listening.

Useful for family conflicts, workplace disputes, and even civic/community dialogue

5. Leader's Reflective Practice

For leaders at any level (from parents to presidents)

Begin each major decision with three reflection questions:

"Am I reacting from fear or responding from clarity?"

"Who will be impacted, and how will they feel?"

"What legacy does this decision create?"
Writing answers down before acting
prevents impulsive, ego-driven decisions.

6. The Global Mirror Exercise

Communities or organizations gather
annually to reflect collectively.

Steps:

Share what has gone well.

Name challenges openly.

Reflect on shared values and
responsibilities.

Decide on one action together that honors
both present needs and future generations.

This scales reflection into a cultural habit.

7. The Forgiveness Practice

On a personal or political level,
forgiveness clears the mirror.

Practice by journaling or saying aloud: "I
release the story that binds me to anger."

Not about erasing harm, but about freeing
energy for new solutions.

At a societal level, forgiveness
commissions can heal divides and open
pathways for peace.

8. Daily Global Reflection Question

Each day, ask:

"If everyone lived the way I am living today — in how I consume, how I treat others, how I lead — what kind of world would we create?"

This practice keeps personal actions tied to global impact.

Key Takeaway

Global challenges can feel overwhelming, but they are, in many ways, mirrors of our own shortcomings. They reflect our fears, our habits, and our blind spots back at us. By practicing reflection individually, collectively, and globally, we can polish the mirror, dissolve illusions of separation, and act from clarity. The science of connection begins here: not with new weapons or new markets, but with a new way of seeing ourselves and one another.

Chapter 16 – The Horizon of Consciousness: Toward a Reflective Civilization

The Neural Net of Humanity

Initially, it started as an experiment. Scientists gathered 500 people in a large hall, each fitted with lightweight neural sensor devices that could track brainwaves in real time. The volunteers were asked to sit quietly, close their eyes, and breathe together.

On the screen at the front of the room, the data streams from all those minds merged into a single shifting image, a living tapestry of consciousness. At first, the patterns were chaotic, jagged, reflecting the nervousness and distraction of the group. But as the minutes passed, something remarkable happened. The waves began to synchronize. Coherence spread through the network, like ripples aligning in a calm pond.

The scientists watched in awe as the image smoothed into harmony. For the first time in history, they weren't just measuring individual minds, they were witnessing a collective state of awareness.

News spread quickly. Soon, gatherings like this were held around the world in temples, universities, and city squares. Each time, the same thing happened: when people came together in reflection, their minds aligned. And each time, participants left with the same feeling: lighter, clearer, more connected.

The implications went far beyond meditation. Global leaders took notice. Before critical summits on climate or peace negotiations, councils began holding collective reflection sessions. With sensors connected, leaders could see not only their own state of mind but the collective coherence of the group. The sessions often started with dissonant jagged patterns of fear, ego, and mistrust. But with guided reflection, the waves began to harmonize. Decisions that once seemed impossible emerged naturally from the shared clarity.

Ordinary citizens joined too. Communities around the globe began hosting weekly reflection circles, both in person and online, linking their neural states into larger networks. Families used the technology to understand each other beyond words. Schools used it to teach children how their minds influenced one another. Businesses discovered that teams with higher coherence made better, more creative decisions.

It wasn't just data. It was a mirror of humanity itself, showing that our minds were never truly separate, only reflecting fragments of the same field.

Of course, there were debates and fears. Would governments misuse the technology for control? Could corporations exploit it for profit? But alongside these risks, something undeniable was unfolding: the recognition that consciousness is not only personal, but collective.

The discovery reshaped how societies defined progress. Nations once proud of their military or economic might now sought recognition for their Reflective Coherence Index, the degree to which their people could live with clarity, compassion, and connection. Citizens no longer asked only, "What's our GDP?" but also, "What's the quality of our shared mind?"

In time, people began to see what sages and mystics across cultures had always hinted: that reflection was not just about polishing one's individual mirror, but about aligning the great mirror of humanity. And when that mirror grew clear, the possibilities were endless.

The Neural Net of Humanity wasn't a machine or an invention in the usual sense. It was a rediscovery of a proof that the wisdom of "we" had always been greater than the noise of "me."

And from that moment on, reflection was no longer seen as a private luxury. It was understood as the foundation of civilization's next step.

For most of history, humanity has measured progress in material terms: wealth, technology, infrastructure, and power. But as our challenges grow more complex and our interdependence becomes undeniable, a new question arises: what if the next horizon of human advancement is not external at all, but internal? What if the true frontier lies in cultivating individual and collective reflection as the foundation of a wiser, more compassionate civilization?

Convergence of Science and Spirit

For centuries, science and spirituality were seen as opposing forces, one rooted in objective observation, the other in inner experience. Yet in the study of

consciousness, their paths are beginning
to converge.

Neuroscience insights into awareness and
perception. Modern neuroscience shows
that perception is not a direct window to
reality, but a constructed reflection. The
brain filters, edits, and interprets raw
information before presenting us with a
picture of the world. Awareness itself, the
ability to notice these processes, is now
being studied as a distinct form of
cognition. When we step back and reflect,
we are literally reshaping neural patterns,
quieting default circuits, and
strengthening networks of clarity and
choice.

Quantum theory and questions of
consciousness. At the same time, physics
has raised profound questions. In the
quantum world, the observer seems to
play a role in shaping outcomes. While
scientists debate the philosophical
implications, the fact remains that
consciousness cannot be easily separated

from the fabric of reality. This mystery has led some researchers to explore whether reflection is not just a private activity but part of a deeper connection between mind and universe.

Parallels with contemplative traditions across cultures. What science is discovering echoes what spiritual traditions have said for millennia. Buddhist texts describe the "mirror mind," Hindu philosophy speaks of Atman as pure awareness, Taoism compares wisdom to still water, and Christian mysticism speaks of divine union through contemplation. The metaphors differ, but the insight is the same: the mind becomes clear when it reflects without distortion, and in that clarity, truth is revealed.

The Possibility of a Science of Consciousness

If reflection is more than a personal habit, if it is the gateway to understanding consciousness itself, then we may be on the edge of a new scientific frontier. Studying reflection not just as psychology but as awareness itself. Traditional psychology measures thoughts, feelings, and behavior. But new approaches are beginning to ask: what is awareness itself? Can it be quantified? Early experiments suggest that the answer is yes, based on brain scans showing meditation-induced changes and physiological markers of reflective states. Collective consciousness research and its implications. Studies of groups meditating together have found ripple effects on stress levels, cooperation, and even crime rates in surrounding areas. Whether through neural synchronization or shared emotional states, science is beginning to explore how reflection may operate collectively. If awareness connects us, then the implications for peace, governance, and global cooperation are enormous.

The potential for reflection as a measurable, teachable capacity. Just as literacy became a cornerstone of civilization, reflective literacy could become the next. Already, schools teach mindfulness, leaders practice reflective pauses, and workplaces adopt structured self-inquiry. The next step may be reflection measured and cultivated as deliberately as physical health, a fitness of the mind and spirit.

Redefining Progress

Progress has long been defined by what we build, consume, and conquer. But reflection invites a deeper measure: not just how much we create, but how clearly we see and how wisely we act.

Beyond material success, clarity is a measure of advancement. Imagine nations ranked not only by economic growth but by levels of compassion, cooperation, and resilience. A society that consumes endlessly but cannot pause to reflect is wealthy but fragile. A society that cultivates clarity is strong in ways GDP cannot measure.

Social well-being as a reflection of inner well-being. Mental health crises, polarization, and environmental collapse are mirrors of a collective inner state clouded by fear and disconnection. When individuals practice reflection, their clarity ripples outward into families, communities, and policies. The well-being of society is inseparable from the well-being of its members' minds.

Imagining a reflective civilization that values wisdom as much as innovation. In a reflective civilization, innovation is not abandoned but guided. Technology serves awareness instead of distraction. Education teaches reflection alongside math and science. Leaders are trained not only in strategy but in stillness. Progress is defined not only by taller buildings or faster machines, but by more profound compassion and wiser choices.

A Glimpse of the Future

What might life look like in a reflective civilization?

Societies are built on reflection. Communities begin each day with silence or a shared story. Diplomacy opens with perspective-taking, not posturing. Families prioritize reflection practices as much as material success. The mirror of the collective mind is polished regularly, reducing distortion in every area of life.

Technology supporting clarity instead of distraction. Devices evolve from stealing attention to guiding awareness. Instead of endless notifications, future tools may gently remind us to pause, breathe, and reflect. AI could act as a mirror, highlighting blind spots and offering compassionate reframing. In this way, innovation amplifies clarity rather than confusion.

Humanity is evolving toward deeper awareness and compassion. Over time, reflection shifts our sense of identity. Instead of seeing ourselves as separate, we begin to recognize the truth echoed across traditions and science alike: we are reflections of one another, and of the universe itself. Compassion becomes natural because the boundary between "me" and "you" begins to dissolve.

Reflection Tools

1. The Daily Horizon Pause

Each morning, before reaching for a phone or starting tasks, take 3–5 minutes of silence.

Ask one guiding question: "What kind of world do I want to reflect today?"

This sets the intention not just for personal success, but for collective contribution.

2. Clarity Journaling

End each day with two reflections:

"Where did I act from clarity today?"

"Where did I act from confusion or fear?"

Over time, this practice reveals patterns and gradually strengthens awareness of choices.

3. Technology as Ally, Not Distraction

Reframe devices into tools for reflection:
Set calendar reminders for "pause and breathe" moments.

Use journaling apps or AI prompts to explore blind spots.

Switch off unnecessary notifications for at least one hour daily.

The goal is to shift technology from stealing attention to supporting clarity.

4. Shared Reflection Practice

Once a week, gather with a friend, family member, or small group.

Share one experience from the week where reflection changed how you responded.

Listen without judgment, noticing common patterns.

This builds collective reflection — the foundation for reflective communities.

5. The Observer Meditation

Sit comfortably, breathe naturally, and watch thoughts as if they were clouds drifting by.

Instead of pushing thoughts away, notice them and say silently: "Not me, not mine, just passing."

Five to ten minutes daily trains the brain to detach from over-identification with thoughts and emotions.

6. The Reflection Compass (Decision Tool)

Before making an important choice, ask: Does this align with my values?

How will this affect others?

How will I feel about this decision in 10 years?

This simple compass shifts decisions from reactive impulses to reflective clarity.

7. Global Perspective Visualization

Once a week, close your eyes and imagine a person on the other side of the world.

Picture their daily struggles, their hopes, their family.

Ask: "What connects me to them?"

This strengthens empathy and reinforces the reality of shared humanity.

8. Reflection-as-Progress Journal

Instead of only tracking goals like income or achievements, track moments of clarity, compassion, and wisdom.

At the end of each month, reflect on growth in awareness.

This helps redefine "progress" on personal terms, aligned with the vision of a reflective civilization.

Key Takeaway

The horizon of consciousness is not a distant fantasy. It is here, now, in every pause before reaction, in every act of self-awareness, in every moment when we choose clarity over confusion. Reflection is the seed, and if nurtured, it can grow into the foundation of a new civilization, one where wisdom is valued as much as innovation, where compassion is seen as strength, and where humanity finally learns to see itself clearly.

Part 3 – The Mirror of the Self

Clearing distortions to see who you truly are

Before we can change the world around us, we must first learn to see ourselves clearly. Most of the pain, frustration, and confusion we experience in life comes not from the events themselves, but from the way we reflect on them and on ourselves. The mirror of the mind not only shapes how we perceive the world, but it also colors our perception of who we are. When that mirror is clouded, we confuse passing thoughts for identity, mistake perfection for worth, and live as if our value depends on the opinions of others. We chase external markers of happiness, trying to fix the reflection in the mirror rather than polishing the mirror itself. The result is stress, self-doubt, and the constant feeling of not being enough.

But when the mirror is clear, everything shifts. You realize you are not your thoughts or feelings. You recognize your inherent worth without needing validation. You stop seeing the world as hostile and instead begin to notice opportunities. You discover that happiness has always been here, waiting in the present. And you find that real growth is not about becoming perfect, but about embracing imperfection and change with clarity.

This part of the book is about that inner journey. Each chapter explores a common distortion of the self, ranging from over-identifying with thoughts to chasing external happiness to being trapped by other people's judgments. We'll look at the science of why these patterns are so convincing, and the practical ways to free ourselves from them.

The goal is not to become someone new, but to finally see yourself clearly beyond stories, beyond illusions, beyond the fog

of self-doubt. When that clarity arrives,
life feels lighter, relationships deepen, and
change becomes easier.

This is where reflection begins: not out
there, but in here.

Chapter 17 – I Am Not My Thoughts and Feelings

The Storm of Emotion

One moment, everything was fine. The day had been ordinary, not great, not terrible, just another stretch of hours filled with routine tasks and small conversations. Then, out of nowhere, the storm arrived. A disagreement, a sharp comment, a glance that felt dismissive, something small, but enough to send a spark into the mind. And just like that, the weather inside changed.

At first, it was just irritation, like a dark cloud sliding in. But irritation didn't stay small. It built. The mind grabbed hold of it, spinning stories: They don't respect me. They always do this. Why can't they listen? Each thought added more fuel, more lightning, more thunder. Within minutes, irritation had transformed into full-blown anger.

The body joined in, too. Shoulders tightened. Heart raced. Breathing grew short and shallow. The storm wasn't just in the head; it was in every muscle, every nerve. Words came out sharper than intended, louder than necessary. The storm demanded release, and release came in the form of arguments, accusations, and defensiveness.

In that moment, anger felt like truth. It painted everything with its colors. The other person wasn't just disagreeing they were the enemy. The situation wasn't just frustrating; it was unfair, intolerable, and impossible to forgive. Under the storm's spell, the world shrank down to a single, distorted reflection: I am angry, and this anger is who I am.

But storms don't last forever. Hours later, the intensity softened. The thoughts slowed down. The body relaxed. The mind looked back on what happened and thought, Why did I say that? Why did it feel so overwhelming? What had felt permanent and all-consuming now seemed temporary, almost fragile. The storm had passed. And with it came regret, confusion, and sometimes shame. This is the trap of emotions when we mistake them for identity. Anger isn't just a passing state; it becomes a defining characteristic of who I am. Sadness isn't just a wave of feeling; it becomes I am broken. Fear isn't just a nervous thought; it becomes a feeling of being powerless. When the storm is raging, it's hard to remember that the weather constantly changes, that no storm lasts forever.

The truth is simple but easy to forget: emotions are not who we are. They are temporary reflections in the mirror of the mind, like clouds passing across the surface of a lake. When the wind stirs the water, the reflection looks jagged and chaotic. But the water itself, just like the deeper self beneath our emotions, remains whole and undisturbed.

Yet most of us live as if the storm is the truth. We fight, withdraw, or spiral into regret because we mistake passing weather for permanent climate. The cost is heavy: strained relationships, words we wish we could take back, and choices made in a fog rather than clarity.

But the storm holds a lesson. It shows us the power of the mirror. When we learn to pause and recognize that emotions are passing visitors, not permanent residents, everything changes. The storm may still come, as storms are a part of life, but we no longer let them define us. We can watch them move across the sky of the mind and know that behind the clouds, the sky is still clear.

Most people live as though every thought and feeling that passes through their mind is who they are. If the thought says, I'm not good enough, they accept it. If the emotion of anger rises, they declare, I am angry. If sadness lingers, they believe, I am broken. These thoughts and feelings can feel so close, so convincing, that it seems impossible to separate them from identity.

But this is the trap. Thoughts and feelings are not the self; they are experiences within the self. They are reflections in the mirror of the mind, and reflections are not the same thing as reality.

The Trap of Identification

The first mistake people make is confusing thoughts with identity. A thought like I'm terrible at this feels true because it arises inside your own head. However, the origin of an idea does not necessarily make it accurate. Thoughts are often based on habit, memory, fear, or conditioning. They show up uninvited, and they don't need your permission. Yet because they appear in your mind, you assume they must reflect you.

Emotions can be just as deceptive. Anger, sadness, anxiety, or fear often feel like permanent states, like who you are in that moment. But emotions are more like storms moving through the atmosphere. They rise, swell, and fade. The fact that they feel overwhelming does not mean they last forever. Think back to times when you were furious, heartbroken, or terrified. Did it stay that way forever? No. Every emotion eventually gives way to something new.

The third trap is the stories we tell ourselves. The mind doesn't just generate raw thoughts; it builds narratives. Because I failed once, I'll always fail. Because someone rejected me, I must be unlovable. Because I feel nervous, I must be weak. These stories act like distorted mirrors. Instead of reflecting reality, they twist it, making temporary experiences appear like permanent truths.

The Science of Inner Reflections

Modern science explains why these patterns feel so convincing. Deep in the brain lies the Default Mode Network (DMN). This system activates when the mind is not focused on an external task.

It's the part of you that daydreams, replays past conversations, and imagines the future. Most importantly, it produces a constant stream of self-talk that runs like background music.

This network is useful for creativity and planning. Still, when it becomes overactive, it leads to rumination, the endless replaying of problems, mistakes, or worries. Rumination keeps the mirror foggy. Instead of reflecting the present moment, the mind loops through old fears and future anxieties.

Psychologists distinguish between rumination and healthy reflection. Healthy reflection asks questions like, What can I learn from this? How do I move forward? Rumination, on the other hand, asks, What's wrong with me? Why did I fail? Why is life unfair? Healthy reflection clears the mirror. Rumination smears it with distortions.

Fortunately, the brain is flexible. Studies show that mindfulness, the practice of noticing thoughts and returning to the present, actually rewires the brain. Regular mindfulness reduces activity in the DMN, which means fewer cycles of overthinking. It also strengthens regions of the brain associated with attention and emotional regulation. In simple terms, mindfulness trains the brain to step back from thoughts and emotions, rather than being consumed by them.

Life Beyond Identification

What happens when you stop believing every thought and emotion? Life feels different. Imagine waking up with the thought, I'm not ready for today. Instead of taking that thought as truth, you notice it. You think, There's the "not ready" thought again. Already, there is space between you and the reflection. The thought loses its grip.

The same is true with emotions. Anxiety can still rise, but instead of being consumed by it, you observe it. You feel the tightness in your chest, the racing thoughts, the quickened breath, but you also know, This is just a passing storm. That awareness alone softens its power. Even the stories begin to shift. Instead of believing the old narrative, "I always mess things up," you recognize it as a story, not a fact. And once you see it as a story, you have the freedom to write a new one.

This clarity changes everything. You become calmer in conflict because you don't confuse someone else's harsh words with your worth. You bounce back from setbacks faster because you don't interpret failure as identity. You experience greater peace because you're no longer trapped inside every storm that passes through.

Living Beyond Thoughts and Feelings

The key is developing the "observer" perspective. Imagine standing on a riverbank, watching thoughts and emotions float by like leaves on the water. Some are pleasant, some are painful, but none of them define the river itself, and none of them define you. You are the observer of the flow, not the leaves themselves.

Practical exercises help strengthen this perspective. One simple method is journaling. Write down a thought or emotion, and then add the phrase, "I notice the thought that…" or "I notice the feeling of…" For example: "I notice the thought that I'm not good enough." This small shift reminds you that you are the one who notices, not the thought itself.

Another practice is the mindfulness pause. Set a timer for just one minute. Close your eyes and focus on your breath. When thoughts or emotions arise, don't fight them, simply label them: "thinking," "feeling," and return to the breath. Over time, this trains your mind to notice without clinging.

As this perspective grows stronger, freedom follows. You begin to realize that thoughts and feelings are not enemies; they're simply reflections. Some are helpful, some are distorted, but none of them define who you are. The truth is bigger, deeper, and calmer than any single reflection.

Reflection Tools

The Noticing Exercise

When a thought or feeling arises, say to yourself: "I notice the thought that…" or "I notice the feeling of…" This creates space between you and the experience.

One-Minute Pause

Set aside one minute a day to focus only on your breath. Each time your mind wanders, gently label it "thinking," "feeling." Return to the breath.

Reframing Stories

When you catch yourself telling a negative story ("I always fail"), write it down. Then rewrite it as: "I sometimes struggle, but I can learn and grow." Watch how the new story shifts your perspective.

Key Takeaway

When the mirror clears, life opens. You no longer live trapped inside passing storms.

Instead, you stand in the clarity of knowing that you are more than what moves through your mind. You are the one who sees.

Chapter 18 – I Am Enough

The Mirror of Comparison

Every time the phone lit up, the world opened like a window into other people's lives. Vacations in places that looked like postcards. Perfectly decorated homes, glowing with the warmth of people who seemed endlessly happy. Photos of bodies shaped by workouts that looked effortless, smiles that seemed natural, successes that came across as inevitable.

At first, scrolling was entertainment. Just a quick break in the day, something to pass a few minutes. But soon, it became something else, a mirror. Each image didn't just show someone else's life; it reflected something missing in their own. Why don't I have that? Why don't I look like that? Why am I not as far ahead as they are?

The comparisons multiplied. One person's promotion made their own job feel stagnant. Another's new relationship made their single life feel empty. Even small things, a friend's dinner out, a cousin's new car, carried the quiet whisper: They're doing better. You're behind.

It didn't matter that reality was different. Their own life had its bright spots, achievements, moments of laughter, and people who cared. But none of that seemed to matter in the face of the mirror that scrolling created. In that reflection, the story was always the same: Not enough. Not yet. Not ever.

The strange part was that the comparisons weren't even accurate. Behind those posts were struggles, doubts, and insecurities that didn't show up on the screen. The curated images were just highlights, not the whole movie. But the mind didn't care. The reflection in the mirror felt real. And once the story of inadequacy started, it fed itself, pulling in every memory of falling short, every moment of embarrassment, every fear of failure. The more they scrolled, the smaller they felt. It was as though their own lives, with all their unique moments, quiet joys, and very real worth, disappeared in the glow of the screen. In its place was only the distorted reflection of "not enough."

The truth is, this trap is everywhere. It's not just social media. It shows up in workplaces, classrooms, families, and friendships. We are constantly shown what others have, what others do, and what others are. And unless we step back, unless we polish the mirror of the mind, we mistake those comparisons for reality. We forget that enough-ness isn't something measured against others, it's something inherent, something that was always there, waiting to be seen.

The story of "not enough" is one of the most common and most convincing illusions people live with. It's the quiet voice that says, I should be further ahead by now. I don't measure up. Everyone else has something I lack. Sometimes it shouts, sometimes it whispers, but either way, it shapes the mirror we use to see ourselves. And once we believe it, we begin living as though it were true.

This chapter is about exposing the lie. Because "not enough" is not a fact, it's a distortion. It comes from conditioning, culture, comparison, and fear. So once we understand its origin, we can begin to reclaim the inherent worth that has been there all along.

The thought creeps in quietly: I should be further along by now. I'm not as good as they are. Something is missing. For many, this thought isn't a rare visitor; it's the background noise of daily life. It shapes choices, drains confidence, and fuels an endless chase for proof of worth.

However, the truth is that "not enough" is not reality. It's a reflection, a distorted one. It's a fog that settles on the mirror of the mind, convincing you that your value is conditional, when in fact it has always been inherent.

The Roots of Inadequacy

The illusion often begins early. As children, many of us were compared to siblings, classmates, or even the version of ourselves that adults wished we would be. A sibling gets praised for grades, and you wonder why you didn't measure up. A classmate excels in sports, and you wonder why you're not "talented." Even with love present, the subtle message can sink in: Your value depends on achievement, approval, or performance.

Culture amplifies the message. From the first day of school, we're ranked, scored, and sorted. Later in life, jobs, promotions, and social status continue the cycle. We learn to measure worth by titles, salaries, or likes on a post.

The media makes it worse. Every day, we're shown polished lives that seem more beautiful, successful, or exciting than our own. The message is everywhere: Happiness is for people who look like this, earn like this, live like this. Even when we know it's curated, our minds still compare.

No wonder "not enough" feels so convincing. It's not just a passing thought; it's been rehearsed for years. The brain mistakes repetition for truth. A thought repeated long enough becomes a belief. And when that belief colors the mirror, every success feels temporary, every joy incomplete, and every failure like confirmation.

The Psychology of Self-Worth

Psychologists have shown that much of this comes down to mindset. Carol Dweck's research distinguishes between a fixed mindset and a growth mindset. With a fixed mindset, worth feels static: I am either good enough or I'm not. Every failure feels like proof of inadequacy. But with a growth mindset, worth is not up for debate; challenges are simply chances to grow. Failure isn't an identity; it's information.

Kristin Neff's research on self-compassion takes this further. She found that when people treat themselves with the same kindness they'd show a friend, resilience and motivation actually increase. Many believe that harshness will push them to succeed. In reality, self-criticism fuels shame and avoidance. Self-compassion fuels courage and persistence. It's like polishing the mirror with kindness instead of scratching it with judgment.

Another shift comes from how we frame failure. When failure is seen as identity (I am a failure), the mirror distorts. However, when failure is viewed as an event (I failed this time), it becomes part of the growth process. This reframing doesn't just protect, but it also strengthens it. Resilient people aren't those who never fail; they're the ones who don't let failure define them.

Life When Enough-ness Is Realized

What happens when the fog of "not enough" lifts? Life feels different. Instead of scanning every situation for proof of inadequacy, you move with quiet confidence. Not arrogance, not the need to shout your value or prove it endlessly, but the calm assurance that you don't need to earn worth.

Daily life changes in subtle but powerful ways. You stop overexplaining to win approval. You stop saying yes when you mean no, just to avoid rejection. You stop chasing validation from people who can never give you enough to satisfy the illusion.

Choices become clearer. You choose work, relationships, and goals based not on what will make you "enough," but on what truly matters to you. Relationships deepen because you're no longer hiding behind masks. Creativity flourishes because fear of judgment doesn't control the process.

Imagine waking up and feeling no pressure to prove yourself before you've even had breakfast. Imagine starting a project not with fear of failure but with curiosity about what you'll discover. Imagine looking at yourself in the mirror and seeing not a list of flaws, but a person already whole. This is what life looks like when "enough" is no longer on trial.

Reclaiming Your Inherent Value

Recognizing enough-ness isn't a one-time epiphany. It's a daily practice of polishing the mirror. Gratitude is one of the most powerful tools. When you pause to notice what you already have, your health, your skills, your relationships, your breath, the focus shifts from lack to sufficiency. Gratitude reveals that life is already abundant, and so are you.

Affirmations, when used with sincerity, can also rewire the story. Instead of forcing positivity, choose simple truths that counter the illusion: I am enough as I am. My worth is not up for debate.

Repetition matters because beliefs are built on repetition.

Comparison can be reframed, too. Instead of seeing someone else's success as proof of your lack, see it as a connection. Their achievement is evidence of what's possible for humans, including you. Their beauty, creativity, or strength doesn't diminish yours. When comparison shifts into connection, inadequacy loses its grip. Over time, these practices build a new identity not one tied to titles, looks, or approval, but one rooted in wholeness and authenticity. The illusion of "not enough" may still arise, but you'll recognize it for what it is: a passing reflection, not the truth.

Reflection Tools

Gratitude Mirror

Each morning, write down three things you appreciate about yourself, not achievements, but qualities: kindness, humor, patience, creativity. This shifts the reflection toward inherent value.

Affirmation Reset

Choose one simple affirmation, like "I am enough as I am." Repeat it three times whenever self-doubt arises. Over time, repetition plants a new default belief.

Comparison to Connection

When envy arises, pause and say: "Their success is proof of possibility, not proof of my inadequacy." Notice how this transforms shrinking into expansion.

Self-Compassion Letter

When you fail or struggle, write a letter to yourself as if you were your best friend. Read it back, and notice how kindness softens shame.

Key Takeaway

The illusion of inadequacy is strong because it has been rehearsed for years. But like any illusion, it weakens when exposed. You were never "not enough." That was only a story, a distorted reflection of reality. Beneath the fog, the truth has been waiting all along: you are enough, and you always were.

Chapter 19 – The World Is Working with Me

The Late-Night Walk

The street was quiet, but the mind was loud. Every sound seemed amplified in the stillness of the night. The rhythm of footsteps echoed against the pavement, and though they matched perfectly with each step taken, the thought crept in: What if someone's following me? Shadows stretched long under dim streetlights, distorting the shapes of trees, fences, and parked cars. One shadow in particular looked unusual, hunched and sharp-edged. The heart skipped. Is that someone waiting? Watching? Muscles tensed, ready to run, and the breath quickened. Stress hormones surged, sharpening every sense but narrowing the field of vision.

As the walk continued, small noises became potential threats. A rustle of leaves? Maybe just wind, or maybe someone hiding. The distant hum of an engine? Maybe nothing, or maybe a car slowing down. The mind, caught in fear's grip, spun stories faster than reality could confirm them. Each ordinary detail of the night turned into a possible danger. Finally, the figure in the shadow came into view. It wasn't a person at all, just a tree branch leaning oddly against a lamppost, its silhouette transformed by darkness into something menacing. The footsteps "behind" turned out to be nothing more than an echo bouncing from nearby buildings. The world had not been hostile. The danger had not been real. The threat existed only in the mirror of the mind.

With each realization, the tension loosened. Shoulders relaxed, breathing slowed, the heart calmed. The street was the same as it had been all along, still, quiet, even safe. What changed was not the world outside, but the reflection inside.

This is how fear works. It narrows perception, primes the body for danger, and turns neutral details into signs of hostility. It paints the mirror with shadows and convinces us they are real. Fear makes the world appear to be working against us, even when it isn't.

The truth, revealed step by step that night, was simpler: the world had not been against the traveler at all. The world had just been itself, silent streets, ordinary trees, a sky full of stars. The reflection of fear had created a hostile story, but reality held no such threat.

Fear has a way of turning trees into monsters and echoes into threats. It narrows the mirror of the mind until all we can see are dangers, even when none exist. But just as the night walk revealed, the world is not always against us. Often, it is our own stress and fear that project hostility onto neutral situations. The challenge and the opportunity are to learn how to polish the mirror so we can see life as it truly is: not an enemy, but an ally.

Fear has a way of shrinking the world. A quiet room feels heavy with judgment. A neutral glance becomes a sign of disapproval. A challenge appears to be a trap, not an opportunity. When stress takes hold, the mind's mirror warps, and the world appears hostile even when it isn't.

But what if the opposite is true? What if, beneath the distortions of fear, life is not working against you but with you? What if every obstacle is an ally in disguise?

The Lens of Fear

Stress changes the way we see. When we feel threatened, the brain's perception narrows. This is useful in survival; if a predator lurks, it's better to focus on the danger than the flowers. Neuroscientists call this tunnel vision of the mind. In stress, attention zooms in on threat signals, while possibilities and neutral details fade to the background.

This narrowing is fueled by what psychologists call negativity bias. For our ancestors, spotting danger was more important than noticing beauty. Missing a threat could be fatal; missing a good view was not. So the brain evolved to give more weight to bad news than good, to criticism over praise, to fear over hope. The problem is that in modern life, the "predators" are usually not life-or-death. They're emails, deadlines, or conversations. Yet the same survival reflex kicks in. A colleague's silence feels like rejection. A delay feels like a disaster. Fear paints neutral situations as hostile, and the reflection looks real.

The Cycle of Defensive Reflection

Once fear distorts the mirror, a cycle begins. We create what psychologists call enemy images, mental villains we cast in roles of opposition. A boss, a partner, a stranger online in our minds, they become obstacles to our peace or success.

This leads to a self-fulfilling prophecy. When we expect opposition, we interpret everything through that lens. The coworker's brief email feels curt, so we respond defensively. They sense the tension, and suddenly, conflict is real. What began as a projection becomes reality.

Stress hormones keep the cycle running. When cortisol floods the system, the body braces for conflict, heart racing, breath shallow, muscles tense. In that state, the mirror fogs even more, and hostility seems everywhere. Fear creates tension, tension creates reaction, and reaction creates the very hostility we were afraid of.

But this cycle can be broken. And when it is, something surprising happens: the world begins to look less like an enemy and more like an ally.

Life Through a Clear Lens

Imagine walking into a room without scanning for a threat. Neutral expressions stay neutral. Silence is not rejection but space. The mind is calm enough to notice the whole picture, the supportive nods, the opportunities hidden in the challenge, the chance for connection where hostility once seemed certain.

This is what life feels like when fear no longer dominates perception. The mirror reflects clearly instead of distorting. Stress doesn't disappear forever, but it no longer controls the story.

Relationships transform. Instead of assuming the worst, you leave room for curiosity. You listen more openly, respond with patience, and discover that most people aren't against you; they're simply caught in their own reflections.

Opportunities expand because what once looked like a closed door reveals itself as an invitation. You begin to see allies where you once imagined enemies.

The difference is not in the world itself but in the lens. When the mirror clears, hostility dissolves into possibility.

Shifting the Reflection to Opportunity

The shift begins with reframing. A challenge is not proof of life's hostility; it's an invitation to grow. The tough project at work is an invitation to develop resilience. The disagreement at home is an invitation to deepen understanding. The setback is an invitation to discover hidden strengths.

Mindfulness and breath are powerful tools here. When fear loops start the racing thoughts, the tight chest, the anxious predictions, a single deep breath interrupts the cycle. It signals the body to calm down, lowering cortisol and widening perception again. The mirror clears, even if only for a moment. That moment is enough to choose a different response.

Building trust in life happens through small acts of courage. Each time you face a situation, fear paints it as hostile and, when you discover it wasn't, trust grows. Step into a difficult conversation and find a resolution instead of conflict. Share an idea and find support instead of ridicule. Each act of courage polishes the mirror, making it easier to see the world as working with you, not against you.

Over time, this becomes a way of life. Challenges stop looking like punishments and start feeling like stepping stones. Fear may still whisper, but it no longer defines the reflection. The truth comes through: the world is not your enemy. The world is your partner in growth.

Reflection Tools

Three-Breath Reset

When fear rises, pause. Inhale slowly for a count of four, hold for two, exhale for six. Repeat three times. This calms the body and clears the lens before reaction takes over.

Reframe the Challenge

Write down a current challenge. Then ask: What invitation does this hold for me?

List at least one skill, strength, or insight that you could grow in.

Enemy Image Flip

Think of someone you see as an obstacle. Write down three neutral or positive reasons for their behavior. This weakens the "villain" story and makes space for new possibilities.

Small Acts of Courage

Choose one small step each day that fear tells you to avoid — share an idea, start a conversation, or take a new path. Notice how often the world responds with support instead of hostility.

Key Takeaway

Fear paints enemies where there are none. Stress shrinks perception until the world looks like opposition. But when the mirror clears, a different truth appears: life is not against you. Life is working with you, shaping you, guiding you, and inviting you forward. The question is not whether the world is on your side. The

question is whether you can see clearly
enough to recognize that it already is.

Chapter 20 – Happiness Is Already Here

The Promotion That Didn't Deliver

For years, everything pointed toward one goal: the promotion. It was the symbol of arrival: the corner office, the new title, the recognition that they had finally "made it." Every late night at the office, every weekend sacrificed, every skipped vacation was justified with the same thought: It will all be worth it when I get there.

The dream was polished in the imagination like a shining trophy. They pictured the day it would finally happen, walking into the new office with confidence, coworkers offering congratulations, and family looking at them with pride. In that vision, happiness wasn't just possible; it was guaranteed. And then, one day, it happened. The promotion letter came. Friends congratulated them. The family celebrated. They were given the bigger paycheck, the new desk, and the office with a window view. For a few days, it felt euphoric like stepping into a long-awaited version of life.

But then the glow faded. The first week turned into the second, and stress crept back in. Emails still piled up. The corner office felt lonely. The new responsibilities came with new worries. That sense of "finally enough" started to slip away, replaced by the same old voice: Now I need to prove I deserve this. Now I need to aim for the next level.

It was confusing at first. How could something they worked so hard for, something they believed would bring lasting happiness, feel so ordinary so quickly? The high was real, but it was temporary. Within weeks, life felt much like it had before, just with a bigger paycheck and a nicer chair.

This is the trap of external happiness. The mind whispers, I'll be happy when I get the promotion, when I buy the house, when I find the relationship, when I reach the next milestone. Sometimes, those milestones really do bring a rush of joy. But the rush doesn't last. The brain adapts, the excitement normalizes, and soon the same cycle begins again.

The truth is, chasing happiness outside ourselves is like chasing the horizon. No matter how far we run, it always moves just a little farther away. The world promises fulfillment through achievement, wealth, and status, but the mirror shows a different story: none of it can sustain happiness on its own.

That doesn't mean promotions, achievements, or possessions are meaningless. They can bring joy, comfort, and pride. But when we mistake them for the source of happiness, we set ourselves up for disappointment. Real happiness is not in the corner office or the next milestone. It's already here, waiting to be noticed in the present moment.

This is the hidden trap of chasing happiness outside ourselves.

Achievements and possessions can spark joy for a moment, but the glow fades, and the old patterns return. It's not because we're broken or ungrateful, it's because the mind is wired to adapt. The real shift comes when we stop chasing the horizon and start noticing what's already here. This chapter is about uncovering that shift and how to move from seeking happiness to actually living it.

The Mirage of External Happiness

Modern culture paints happiness as something just out of reach.

Advertisements tell us that joy can be bought with the right clothes, the right car, the right vacation. Social media adds fuel by displaying highlight reels of other people's lives, convincing us that we are behind. The story is clear: I'll be happy when I have more, look better, or achieve what others admire.

But this is a mirage. Even when we get what we want, the satisfaction doesn't last. Psychologists call this hedonic adaptation. It means that after an initial burst of happiness, we quickly return to our baseline mood. The new phone, the bigger house, the dream vacation, they bring a temporary lift, but soon they become the new normal. The mind adapts, and the hunger returns.

This endless pursuit is exhausting. The "I'll be happy when…" story keeps us chasing the horizon. And like the horizon, it never arrives.

The Science of Fulfillment

So if external milestones don't bring lasting happiness, what does? Research points to a different path: fulfillment comes more from the inside than the outside.

Studies on intrinsic vs. extrinsic motivation show that when people focus on extrinsic goals, such as money, fame, and image, they may achieve them, but their happiness quickly fades. Intrinsic goals like growth, relationships, creativity, and purpose are strongly tied to long-term well-being. These are things that matter not because others reward them, but because they feel meaningful to us.

Positive psychology research also highlights two key ingredients: meaning and flow. Meaning comes from feeling that what we do has purpose, that our life connects to something larger than ourselves. Flow comes from being so absorbed in an activity that we lose track of time, painting, running, coding, cooking, playing music, anything that brings deep engagement. Together, they create a happiness that isn't fleeting, but sustaining.

And then there's gratitude. Study after study shows that gratitude is one of the most reliable practices for boosting happiness. Why? Because it shifts the mind's focus from what is missing to what is present. Gratitude clears the mirror, allowing us to see abundance instead of lack.

The science is clear: happiness is less about what happens to us and more about how we engage with what's already here.

Life When Happiness Is Within

When you stop chasing happiness "out there," life takes on a different quality. Joy becomes easier to find in ordinary moments, a conversation with a friend, a quiet morning walk, the smell of coffee, the laughter of a child. These moments aren't flashy, but they're real.

Relationships can shift as well. When you no longer need other people to prove your worth, you connect with them more authentically. Instead of using relationships as validation, you use them as opportunities for genuine connection. The pressure lifts, and love becomes freer.

Work changes as well. Instead of being only about titles or salaries, it becomes about contribution, creativity, and growth. Decisions feel lighter when they aren't tied to proving yourself.

Perhaps most importantly, life develops a steady energy. When happiness is within, you no longer swing wildly between highs and lows depending on circumstances. The mirror reflects calm. Even when life is difficult, a quiet contentment remains. This doesn't mean every day is easy, but it means your joy is less fragile. It doesn't depend on constant external proof.

Returning Happiness to the Present

The path to this kind of happiness isn't complicated, but it does require practice. It begins with noticing joy in small moments. Instead of racing past the present to the next goal, pause to savor what is here: the warmth of sunlight, the taste of food, the sound of music. Small joys accumulate, creating a foundation of contentment.

Shifting from seeking to savoring is the key. Seeking says, I'll be happy when something happens in the future. Savoring says, I am happy now. Seeking keeps the mirror fogged with longing. Savoring clears it with presence.

Finally, aligning daily life with values creates deeper fulfillment. When your actions reflect what truly matters to you (kindness, creativity, service, learning, family, and freedom), happiness flows naturally. External circumstances may shift, but alignment remains steady.

Reflection Tools

Gratitude Journal

Each night, write down three things you're grateful for. Make them specific and small: a kind word, a meal you enjoyed, a laugh you shared. Over time, this rewires your brain to notice abundance instead of lack.

Savoring Practice

Choose one everyday activity, such as drinking tea, walking, or cooking, and

practice being fully present with it.
Engage your senses. Let it be enough in
that moment.

Values CheckIn

Once a week, ask: Did I live in alignment
with my values today? If not, what small
adjustment can I make tomorrow? This
shifts the focus from outcomes to
authenticity.

The "Already Here" Reminder

When you catch yourself thinking, I'll be
happy when something happens in the
future. Pause and ask: What can I
appreciate right now? This interrupts the
chase and brings happiness back to the
present.

Key Takeaway

The truth is simple but radical: happiness
is not a destination. It's not hiding in the
next achievement, the next purchase, or
the next milestone. It is already here,
waiting to be seen in the present moment.
When you stop chasing and start noticing,
the mirror clears. And in that clarity, you

discover what was true all along: joy is not out there. Joy is right here, now.

Chapter 21 – I Am Perfect

The Book That Never Felt Ready

The idea came like a spark, a book that could inspire, help, or maybe even change lives. At first, it felt exciting, almost urgent. Pages began to fill quickly. Characters or chapters took shape. There was momentum, energy, even joy.

But then came the edits.

A sentence didn't sound quite right. A paragraph needed to be sharper. The introduction should be rewritten. That one chapter? It wasn't good enough yet. Each time they fixed one part, another flaw seemed to appear. The work became less about expressing ideas and more about hiding imperfections.

Days turned into weeks. Weeks turned into months. The book grew heavier, not because of its content, but because of the pressure to make it flawless. Each time they opened the document, the inner critic whispered: It's not ready. You're not ready. People will see the mistakes. Perfection became the goal, but perfection has no finish line. Every edit revealed new things to change. Every improvement made the gap between vision and reality feel bigger. The more they polished, the further away "done" seemed to be. Eventually, the excitement that started the project turned into dread. They stopped writing as often. They avoided opening the file altogether. What once felt like a bright possibility now felt like a burden they couldn't carry.

And in the end, the book never left the computer. It sat there, half-written or endlessly revised, waiting for the moment when it would finally be "good enough." That moment never came.

The tragedy wasn't that the book had flaws; all books do. The tragedy was that the world never got to see it at all. Readers who might have been helped, moved, or inspired never had the chance because the pursuit of perfection smothered progress.

This is the hidden cost of perfectionism. It promises excellence but often delivers paralysis. It convinces us that if we keep polishing, we'll finally feel safe to share. However, the truth is that perfection is a moving target. No matter how close you get, it always moves a little farther away. The book didn't need to be flawless. It only needed to be finished, shared, alive in the hands of readers. Its imperfections might have made it more human, more relatable, more impactful. Instead, it stayed locked in the private prison of "not good enough."

Perfectionism pretends to protect us from failure or rejection. But what it really does is guarantee that growth, connection, and opportunity never happen. And nowhere is that more evident than in the story of the book that never left the page.

This is the illusion of perfection: it convinces us that flawless work or a flawless self will finally bring safety, approval, or peace. But instead of freeing us, it traps us. Perfection is a distorted reflection, one that keeps moving, no matter how hard we chase it. The truth is, growth, creativity, and connection don't come from being flawless. They come from being real. This chapter is about breaking free from the perfection trap and discovering the freedom of imperfection.

The Illusion of the Perfect Self

Perfectionism doesn't appear out of nowhere. It is fed by cultural pressures that bombard us daily. Advertisements show flawless faces and bodies, suggesting that beauty equals worth. Social media shows curated lives, convincing us that everyone else is happier, richer, and more successful. Schools and workplaces reward achievements while overlooking effort, reinforcing the belief that value comes only from results.

Comparison fuels the fire. When we measure ourselves against filtered versions of other people's lives, it's easy to feel behind. Someone else seems more productive, more talented, more attractive, more everything. In that distorted reflection, perfection feels like the only way to catch up.

But perfection is a moving target. Each time you get close, the standard shifts. You finish the project, and instead of feeling proud, you notice the flaws. You achieve the goal, and instead of celebrating, you immediately set a higher one. Perfection isn't a destination; it's a treadmill. And no matter how hard you run, you never arrive.

The Psychological Cost of Perfectionism

This endless pursuit takes a heavy toll. Research shows that perfectionism is closely linked to anxiety, depression, procrastination, and burnout. The fear of not being flawless can stop people from starting, just as it can prevent them from finishing. It convinces us that unless something is perfect, it isn't worth doing at all.

Beneath perfectionism lies a fear of failure, a fear of rejection, and a fear of being exposed as "not enough." Instead of letting us grow, it locks us into playing it safe. Instead of encouraging boldness, it keeps us trapped in hesitation.

Yet there is another way. Kristin Neff's research on self-compassion reveals that individuals who treat themselves with kindness, particularly when they fall short, are more resilient and motivated in the long run. Counterintuitively, letting go of the need to be perfect doesn't lower standards; it makes achievement healthier and more sustainable. Compassion clears the mirror where perfectionism fogs it.

Life When Imperfection Is Embraced

What happens when you step off the treadmill of perfection? Life opens. Progress becomes more valuable than polish. Done becomes better than perfect. Suddenly, finishing the book, the project, or the idea matters more than keeping it flawless.

Authenticity shines. When you drop the mask of "having it all together," you make room for creativity and honesty. Mistakes aren't shameful; they're part of the process. Many of the most innovative ideas in history were born from trial and error, not flawless execution.

Relationships deepen, too. People connect to your humanity, not your perfection. Think about it: who do you feel closer to, the friend who admits their struggles, or the one who pretends everything is fine? When you allow flaws to show, you invite others to do the same. The result is trust, vulnerability, and connection.

Life with imperfection embraced isn't sloppy or careless; it's real. It's living with the understanding that growth comes through practice, that beauty often hides

in the cracks, and that progress matters
more than polish.

Growing Through Imperfection

Growth doesn't require perfection. In fact,
it requires the opposite. Mistakes are the
raw material of learning. Each time you
fail, you gather information. Each time
you stumble, you discover resilience. The
mirror clears when you see mistakes not
as proof of inadequacy but as stepping
stones.

Practical exercises can help shift this
perspective. One powerful approach is
reframing mistakes. Instead of saying, "I
failed," you can say, "I learned." Instead
of asking, "What went wrong?" you can
ask, "What can this teach me?" This
reframing turns imperfection into
progress.

Another practice is deliberately choosing
progress over perfection. Set a small,
achievable goal and allow yourself to
complete it even if it's not flawless.
Publish the blog post, share the artwork,

and finish the draft. Each time you choose progress, you weaken perfectionism's grip.

Finally, celebrate small, imperfect steps. Notice the courage it takes to try, the growth that happens even when results are messy. Over time, these celebrations rewire your brain to associate imperfection with possibility, not shame.

Reflection Tools

The 80% Rule

Commit to finishing projects when they feel 80% "good enough." The final 20% of polishing often causes paralysis. Better to share imperfect progress than to hide unfinished work.

Mistake Reframe Journal

Each time you make a mistake, write it down. Then write what it taught you. Over time, you'll see mistakes as proof of growth, not weakness.

Imperfect Action Challenge

Choose one area of your life where perfectionism holds you back. Take one imperfect action this week — send the email, share the draft, speak up in the meeting — and note how it feels.

Celebrate the Cracks

List three "imperfections" in yourself that have brought unexpected strength or connection. For example: "My sensitivity makes me empathetic," or "My awkwardness makes people laugh and feel comfortable."

Key Takeaway

Perfection is not the truth of who you are. It's a distorted reflection, one that keeps you running without rest. The reality is simpler and more powerful: growth, creativity, and connection live in imperfection. When you embrace flaws as part of the process, the mirror clears, and you discover that you don't need to be flawless to be whole.

You don't need to chase perfection. You already have everything you need to grow, connect, and live fully.

Imperfection is not failure; it is the path.

Chapter 22 – I Am Not What Others Think of Me

The Post That Wasn't Good Enough

They had taken the picture earlier that day, a small but meaningful moment they wanted to share with others. Nothing earth-shattering, just a slice of life. The kind of thing that, years ago, would have stayed in a photo album or a journal. But now, with social media, it felt natural to share it with friends, maybe even strangers, as a way of saying, This is me. This is what I value.

They opened the app, uploaded the photo, and began to type a caption. The first draft felt too long. Delete. The second felt too short. Rewrite. They added an emoji, then deleted it. They tried to make it funny, then serious, then inspirational. Each version seemed off.

Minutes ticked by. Their finger hovered over "post," but something inside pulled back. A thought whispered: What if nobody likes it? What if people think it's silly? What if I come across as trying too hard?

They switched the filter, brightening the colors. That looked better, but now it felt fake. They put the original back, but then worried it was too plain. Back and forth, they wrestled with tiny details that had nothing to do with the original moment they wanted to capture.

By now, twenty minutes had passed. The photo wasn't joyful anymore; it was stressful. The caption wasn't a reflection of their voice; it was a script written for an invisible audience. The post had become less about sharing something real and more about avoiding imagined judgment.

Finally, with a sigh, they closed the app. No photo. No caption. No post.

At first, they told themselves it was no big deal. But later, the thought returned: Why did I stop myself? Why does other people's reaction matter so much?

It wasn't about the post at all. It was about the mirror; they were looking into the mirror of other people's opinions. In that mirror, their worth was tied to approval, to likes, to comments. The simple act of sharing something meaningful had been hijacked by fear of what others might think.

The tragedy wasn't that the world never saw the photo. The tragedy was that their authentic self-expression had been silenced. They had something real to share, but it was buried under the weight of imagined criticism.

This is how living in the reflections of others works. It makes us second-guess, hold back, and measure our value through applause or silence. However, the truth is that other people's reactions are merely reflections of their own state of mind, not proof of our worth. And when we give those reflections too much power, we stop living for ourselves.

The post that never got shared shows how deeply we can become trapped in the reflections of others. A moment of self-expression becomes paralyzed by imagined reactions. Approval feels like oxygen, and without it, we suffocate. But this is not who we are. We are not the number of likes, the applause, or the criticisms that come our way. We are more than the reflections that bounce back at us.

Living in the Reflections of Others

From the time we're children, we're conditioned to tie our worth to approval. Parents beam when we get good grades. Teachers reward gold stars. Coaches cheer when we score. Approval feels good; it means we belong, we're accepted, we're safe. Disapproval, on the other hand, feels threatening. It triggers fear of exclusion, punishment, or rejection.

This conditioning doesn't fade as we grow older. Instead, it morphs. In adulthood, it shows up as chasing promotions, compliments, or recognition. In today's world, it's amplified by social media. Platforms built on likes, shares, and comments keep us hooked, training our brains to crave validation.

Comparison becomes constant against coworkers, friends, influencers, and even strangers we've never met.

Why is this so powerful? Because external validation is addictive. Each like, compliment, or nod of approval gives us a small hit of dopamine, the brain's "feel-good" chemical. It rewards us for seeking more. The problem is, like any high, it fades quickly. And so we go back again, posting, performing, and perfecting, hoping for another hit of approval.

But chasing validation comes at a cost. When self-worth is tied to others' opinions, it becomes fragile. One critical comment can shatter confidence. One moment of being overlooked can feel like failure. Life becomes about performing for others instead of living from within.

The Science of Social Mirrors

Psychologists describe this phenomenon through social identity theory. We define ourselves not just as individuals, but as members of groups: family, culture, profession, community. Belonging feels safe, but it can also pressure us to mold ourselves into what others expect.

Self-esteem, too, often becomes externally anchored. Research shows that when self-worth depends on outside approval, people experience more stress, anxiety, and depression. They ride an emotional rollercoaster based on the reactions of others.

Shame plays a central role. Shame researcher Brené Brown describes it as "the intensely painful feeling that we are unworthy of love and belonging." When we live for others' judgments, we live in fear of shame. We hide our flaws, censor our voices, and pretend to be what we think others want us to be. But the cost of hiding is authenticity.

Science is clear: the more our worth depends on external validation, the less resilient and happy we are. True self-worth, on the other hand, grows from the inside out.

Life When Judgment Loses Its Power

Imagine what life feels like when other people's opinions no longer control you. Confidence doesn't come from applause, but from authenticity. You can share your ideas, speak your truth, or show your creativity without needing everyone to approve.

This doesn't mean you stop caring about people. It means you stop confusing their reflections with your identity. Someone else's praise or criticism is more about them, their mood, their perspective, their state of mind, than it is about you. When you see it this way, their words lose power over your sense of self.

Life opens up when judgment loses its grip. You take more risks, not because you're reckless, but because you're free. You love more deeply because you're not constantly guarding yourself against rejection. You create more boldly, because you're not paralyzed by what "they" will think.

Relationships also shift. Instead of trying to earn approval, you show up as you are. This invites deeper connection, because authenticity is magnetic. When you stop hiding behind the mask of perfection, others feel safe to drop their masks too. This is the freedom of living beyond the judgment mirror: the clarity to see yourself as more than other people's reflections.

Reflection Tools

Breaking free from external validation doesn't happen overnight. But with practice, you can build a self-identity rooted in reflection, not reaction. Here are some ways to start:

1. Detach from criticism and praise. When someone criticizes you, pause before reacting. Ask: Is this about me, or about them? Often, criticism reflects the critic's stress, fear, or expectations. Praise works the same way. Instead of basing your worth on it, appreciate it without needing it.

2. Reframe other people's words. Instead of thinking, They rejected me, try: They weren't in a place to receive what I offered. Instead of, they think I'm not good enough, try: They are showing me their lens, not my truth. This simple shift breaks the illusion that others' reflections define you.

3. Practice authenticity in small ways. Start by expressing yourself in low-stakes situations. Share a thought without editing it to be perfect. Wear what feels good to you, not what you think others expect. Each act of authenticity strengthens the muscle of self-trust.

4. Anchor self-worth in reflection, not reaction.

Spend time journaling about who you are beyond roles, achievements, or approval. Ask: What matters to me? What do I value? What do I want to express? These reflections create an identity that is internal, not external.

5. Build shame resilience.

Shame loses power when exposed. Share your struggles with a trusted friend or therapist. Speaking what you fear makes it less heavy. Vulnerability turns shame into connection.

The mirror of others' judgments will always exist. People will always have opinions, and some of them will sting. But you don't have to live inside that mirror. When you stop chasing approval, you reclaim freedom.

Key Takeaway

You are not what others think of you. You are not their compliments or their criticisms. You are not their likes, shares, or silence. You are the one looking in the mirror, not the reflection that bounces back. And when you live from that truth, your life becomes less about performing and more about being.

Chapter 23 – Change Is Easy

The Gym Membership That Gathered Dust

Every January, it started the same way. A surge of motivation, a brand-new resolution, and the swipe of a card for a shiny new gym membership. This was going to be the year. The year of discipline. The year of transformation. The year they would finally become the person they wanted to see in the mirror. The first week went well. They showed up, pushed through the soreness, and even felt a little proud. The second week got harder. Life's usual demands: long workdays, family obligations, the comfort of staying home, began to tug at them. By the third week, workouts felt more like a chore than a choice. And by February, the gym card was sitting untouched in a drawer, buried under papers and receipts.

The cycle repeated year after year. Each December brought regret about wasted money and missed goals. Each January brought fresh determination. But by spring, they always ended up in the same place, frustrated, ashamed, and convinced: Maybe I just can't change. That belief that I can't change became heavier than the unused gym equipment. It wasn't just about fitness anymore. It seeped into other areas of life. Why start the side project? Why learn a new skill? Why try to shift old habits? The voice in their head had an answer ready: You've already proven you can't stick to it. Why set yourself up for failure again?

The truth, though, wasn't that change was impossible. The problem was how they approached it. Each time, they tried to leap into transformation with an all-or-nothing approach. Go to the gym five days a week. Change the diet overnight. Establish new habits with sheer willpower. But willpower fades, and when

the big leap didn't land, they told
themselves they had failed.

What they didn't realize was that change
rarely happens in giant leaps. It happens
in small, steady steps. Neuroscience
shows that habits are built through
repetition, not intensity. A ten-minute
walk every day can be more powerful
than a two-hour workout once a month.
One glass of water in the morning can
help create momentum toward healthier
choices throughout the day. Tiny actions,
repeated consistently, rewire the brain and
prove to the self: I can change.

The gym membership wasn't the problem.
The belief that change had to be perfect
from the start was. And each time they
gave up, the belief that they were "stuck"
grew stronger.

The truth is, no one is fixed. The brain is
built to grow, adapt, and rewire itself.
Change isn't about being flawless; it's
about taking small steps in the right

direction and celebrating progress along the way.

The gym card in the drawer wasn't evidence of failure. It was evidence of a misunderstanding: that change has to be massive, dramatic, and immediate. In reality, change is simple, not always easy, but always possible when we stop chasing perfection and start practicing consistency.

The unused gym membership wasn't proof of failure; it was proof of a misunderstanding. Change had always seemed like a giant mountain that required willpower, sacrifice, and flawless discipline to climb. But in reality, change is less about force and more about a state of mind. It's about how we see ourselves, the stories we believe, and the reflections we trust. When the mirror of the mind is clear, change stops looking impossible and starts looking natural.

The Illusion of Being Fixed

One of the most common misconceptions people carry is the belief that they are "fixed." Thoughts like "I'm just not a disciplined person," or "I'll never change," become part of the identity. They feel permanent, like a label etched in stone.

Neuroscience explains why change feels so hard. The brain is designed for efficiency. Habits form because repeated actions carve pathways in the brain, making them automatic. That's why it's easy to repeat the same routines; the neural "roads" are already paved. Trying to change feels like hacking through an overgrown forest compared to strolling down a familiar highway.

Fear of failure makes it worse. Every attempt that didn't stick gets stored in memory as "proof" that change is impossible. When you've tried to quit smoking five times or started diets that fell apart, the brain holds onto those failures and whispers: Why bother trying again? The illusion of being fixed grows stronger, not because it's true, but because the evidence feels overwhelming.

But illusions lose power when the mirror clears. And science shows that change is always possible.

The Science of Transformation

At the heart of transformation is neuroplasticity, the brain's ability to rewire itself. New habits literally create new connections in the brain. It doesn't matter how old you are or how many times you've failed; as long as you keep practicing, your brain keeps adapting. Change isn't blocked by age or past mistakes. It's built into your biology.

Psychologist Carol Dweck's research on the growth mindset shows why perspective matters so much. When people believe abilities are fixed, they avoid challenges and give up quickly. But when they believe growth is possible, they take risks, persist longer, and learn from setbacks. The difference isn't talent,

it's mindset. Change is not an event; it's a
process.

Behavioral science adds another layer:
small, consistent steps work better than
massive overhauls. Habits compound like
interest. Flossing one tooth can lead to
flossing them all. Putting on workout
shoes often leads to exercise. Drinking
one glass of water can trigger healthier
choices throughout the day. These "micro-
habits" bypass the brain's resistance
because they're too small to trigger fear.
Over time, they snowball into
transformation.

The science is clear: change isn't about
dramatic leaps. It's about rewiring the
brain with steady practice and a mindset
that says, I can grow.

Life When Change Feels Natural

Imagine living with the quiet confidence
that growth is always possible. Instead of
labeling yourself as stuck, you see
yourself as evolving. Challenges stop
looking like walls and start looking like

doors. Every attempt, even the messy ones, becomes proof of progress.

Daily life becomes lighter. Instead of dreading big goals, you focus on small, doable steps. You don't have to conquer everything at once; you just have to keep moving. A ten-minute walk, a single journal entry, one mindful breath each step is evidence that you are changing.

Imperfection no longer feels like failure. Missing a workout doesn't mean you've failed at health. Snapping in frustration doesn't mean you're incapable of patience. Every stumble is part of the process, not the end of it. Progress, not perfection, becomes the mirror you trust.

And there's joy in noticing progress. The first time you realize you no longer react the same way to stress. The moment you see that a habit that once felt forced now feels natural. The pride of looking back and realizing that consistent, imperfect steps added up to real transformation.

This is what life looks like when change is embraced as natural: lighter, steadier, and filled with possibilities.

Reflection Tools

So, how do you step into this perspective? Here are some practical ways to open the door:

1. Start with micro-habits.

Pick the smallest possible version of a habit and repeat it consistently. Write one sentence a day instead of aiming for a whole chapter. Do one push-up instead of a full workout. These tiny steps sneak past resistance and build momentum.

2. Reframe failure.

Instead of asking, "Did I succeed or fail?" ask, "What did I learn?" Every attempt becomes valuable. The more you practice this, the less intimidating the change will feel.

3. Use patience as a tool.

Change takes time because the brain needs repetition to build new pathways. Be patient with yourself, the way you'd

be patient with a child learning to walk.
Every step, even the wobbly ones, counts.

4. Cultivate the right state of mind.

Change feels impossible when you're
stressed and reactive. But in a calm,
reflective state, you see opportunities
more clearly. Practices like breathing
exercises, meditation, or simply pausing
before reacting can create the mental
space where change feels possible.

5. Celebrate small wins.

Each time you keep a promise to yourself,
celebrate it. Smile, write it down, or say it
out loud. These small celebrations
reinforce the identity of someone who can
change.

Key Takeaway

The truth is, change is not a battle to be
won. It's a process to be lived. The
illusion of being fixed disappears when
you realize the brain is designed to grow.
The fear of failure loses power when you
reframe mistakes as learning. The burden
of perfection lifts when you celebrate

progress instead of waiting for flawless results.

Change is not far away. It's not hiding in some distant future. It begins here, in this moment, with the smallest of steps. And when you take those steps, the mirror of the mind reflects something new: not a person stuck in place, but a person capable of endless transformation.

Chapter 24 – Beyond the Surface

The Constant Scroll

The phone buzzed before the sun rose. Without thinking, they reached for it, eyes still heavy with sleep. A quick scroll through emails, then the news, then social media. Ten minutes later, their body was still in bed, but their mind was already running headlines, to-do lists, and opinions from strangers. The day hadn't even begun, and already the mirror of their mind felt smudged.

By mid-morning, they were bouncing between tasks. Work, messages, and notifications. Each buzz pulled their attention like a magnet. They found themselves rereading the same email three times, struggling to focus, their brain split between too many tabs. Lunch came and went, barely noticed.

In the afternoon, fatigue set in. Not the kind of tired that comes from physical work, but the dull ache of overstimulation. Their thoughts felt crowded. Every decision seemed more complicated than it should be. A colleague's tone in a meeting felt sharper than it was. A small mistake felt bigger than it really was. The mirror of their mind no longer reflected reality; it reflected stress.

By evening, even simple interactions at home felt tense. Family conversations blurred into background noise while the phone filled the silence. Scrolling brought brief bursts of distraction, but no real rest. The mind was foggy, restless, unsettled. This was life on autopilot pulled in a hundred directions, overfed with information but undernourished in clarity. The reflection in the mental mirror was distorted, blurred by stress, distraction, and noise. And yet, it felt normal. This was how most days went.

Until one morning, something shifted. Instead of reaching for the phone, they stayed still. Just ten minutes. At first, it was uncomfortable; the silence felt strange, and the urge to grab the phone was strong. But slowly, they noticed their breath. The way the morning light came through the window. The sound of birds outside. The mind wasn't perfectly calm, but it wasn't as fogged as usual.

That evening, they tried journaling. Nothing fancy, just a few lines about the day. At first, the words felt clumsy, like they weren't saying anything important. But as the days passed, the practice began to clear things out. Thoughts that once swirled endlessly found their way onto paper, leaving space behind them.

Over time, the difference became clear. On the mornings with silence, the day felt lighter. Decisions came more easily. Small frustrations didn't hit as hard. The mind, once cluttered like a messy desk, began to feel orderly, calmer, and easier to work with.

They realized something important: the mirror of the mind doesn't polish itself. If left unattended, it clouds over with stress, distraction, and fear. But with simple practices like silence, journaling, and mindful pauses, it can be polished. And when the mirror clears, life looks different. Clearer. Kinder. More real.

The day of constant scrolling revealed something important: the mind's mirror doesn't stay clear on its own. Stress, distractions, and old patterns can cloud our perception, causing us to no longer see reality as it truly is. Instead, we see through filters of fatigue, judgment, and fear. But clarity is always possible. With simple practices, the mirror of the mind can be polished, and when it is, life looks different.

Why the Mirror Gets Clouded

Modern life moves faster than the mind was designed to process. Notifications, emails, deadlines, and social media flood us with more information in a single day than past generations absorbed in a week. Each piece of input leaves a trace, and the accumulation smudges the mirror of awareness. That's why constant stimulation leaves us restless, distracted, and anxious.

But it isn't just overstimulation that clouds the mirror. Old wounds and unexamined beliefs leave their marks as well. Maybe it's a comment from childhood: "You're not smart enough." Or a past failure that still whispers, "Don't try again." These unchallenged stories become smudges, coloring how we see ourselves and the world.

Judgment and fear blur clarity even further. When we approach situations expecting the worst, we distort what we see. A neutral comment sounds like criticism. A small mistake feels like a disaster. Instead of reflecting reality, the mind projects fears onto the mirror.

The good news is, these distortions are not permanent. Just as a physical mirror can be wiped clean, the mind can be polished.

The Practice of Polishing

Polishing the mirror of the mind doesn't require special equipment or years of training. It begins with simple, everyday practices that clear away the fog.

Mindfulness is the daily act of noticing. Each time you bring attention to your breath, your body, or the present moment, you're polishing. You're stepping out of autopilot and choosing clarity. Over time, mindfulness builds awareness like a muscle, making it easier to see thoughts and emotions as passing events instead of fixed truths.

Journaling and self-inquiry offer another tool. Writing down your thoughts slows them down, giving you the chance to see them clearly. A looping worry that feels overwhelming in your head often shrinks once it's on paper. Journaling isn't about perfect sentences; it's about creating space. When you ask questions like, "Is this thought true?" or "What am I afraid of here?" you shine light on old beliefs that may no longer serve you.

Silence and stillness act as natural cleansers. Taking time each day, even just a few minutes, to sit quietly resets the mind. In stillness, the noise settles, and clarity returns. It may feel uncomfortable at first, but silence is not emptiness; it is the space where reflection happens.

Each of these practices is like wiping the mirror clean. They don't erase challenges, but they let you see them for what they are, without the distortions of distraction, old wounds, or fear.

Life with a Clear Mirror

What does life look like when the mirror is polished? First, it feels calm. Not the absence of activity, but a steady presence underneath it. Stressful moments still happen, but they no longer throw you into chaos. The mirror reflects events clearly, without magnifying them into crises. Clarity also brings compassion. When you're not clouded by your own judgments and fears, you see others more accurately. Instead of reacting defensively, you notice what they're going through. Anger softens into understanding. Misunderstandings resolve more easily, and focus sharpens. With a clear mirror, decision-making becomes simpler. Instead of spinning in circles, you can see the next step. Choices aren't perfect, but they're grounded. You're less likely to overreact to temporary emotions or be swayed by outside noise.

The ripple effects are powerful. Relationships deepen because you're more present. Work feels lighter because you can focus without constant distraction. Purpose emerges more clearly because you're not caught in the fog of fear or comparison. With clarity, life

doesn't just look different, it feels
different.

Living with a Clear Mirror

Polishing the mirror isn't a one-time act.
It's an ongoing practice, like brushing
your teeth or cleaning your home. Life
will always bring stress, wounds, and
distractions, but you can keep clearing
them away.

Living with a clear mirror means seeing
reality as it is, not as fear paints it. A
challenge becomes a challenge, not a
catastrophe. A mistake becomes a moment
of learning, not proof of inadequacy. A
person's criticism becomes a reflection of
their state, not your worth.

It also means relating to others with
compassion and openness. When your
mirror is clear, you no longer need to
project judgment or defensiveness. You
can listen without fear, connect without
masks, and love without needing
perfection.

Most importantly, living with a clear mirror allows you to make choices rooted in clarity, not distortion. Instead of reacting from stress or fear, you respond with presence. You choose actions that align with your values, not with old habits or external pressure.

Polishing the mirror doesn't make life flawless, but it makes life real. And when life is seen clearly, it becomes far easier to live with peace, purpose, and connection.

Reflection Tools

One-Minute Mindfulness

Pause once an hour to take three deep breaths. Notice what you see, hear, and feel in the present moment. This tiny act wipes away a layer of distraction.

Evening Journal Sweep

Before bed, write down the three thoughts that weighed on you most that day. Ask yourself: Is this true? Is this helpful? Can I let it go?

Five Minutes of Silence

Sit without distractions for five minutes a day. No phone, no tasks. Let your thoughts settle like sand in water.

Reframe the Fog

When you catch yourself lost in judgment or fear, imagine it as a smudge on a mirror. Ask: What would this situation look like if the mirror were clean?

Key Takeaway

The mind's mirror is constantly reflecting. The question is whether it's showing a distorted image or a clear one. Stress, wounds, and fear will always try to cloud it, but you have the tools to keep it clear. And when the mirror shines, you see life not as you fear it to be, but as it truly is, full of possibility, connection, and calm.

Chapter 25 – Seeing Yourself Clearly

The Argument That Wasn't Worth Having

The words landed sharply, like stones thrown across the table. A colleague's tone was clipped, critical, maybe even unfair. In the past, this would have been enough to spark an instant reaction. The old version of them would have snapped back, defending every detail, fueling a battle of egos that left both sides drained and nothing resolved.

But this time, something different happened.

The familiar surge of defensiveness rose, accompanied by a quickened heartbeat, tension in the chest, and the urge to speak over the other person. Yet instead of giving in to the reflex, they paused. A single breath. A moment of awareness.

In that pause, they saw the situation more clearly. Their colleague wasn't an enemy. They weren't attacking out of malice. The sharp words were likely a reflection of stress, frustration, or fear. Maybe it had nothing to do with them at all.

That simple realization shifted everything. Instead of throwing more fuel on the fire, they listened. Not passively, not with resentment, but with curiosity. They asked a question instead of launching a defense. The conversation slowed. The colleague's tone softened. What could have turned into a spiraling argument became an honest exchange of views.

Later, reflecting on the moment, they realized how different it felt to live with clarity instead of distortion. Before, the mirror of their mind had been clouded with judgment and fear. Every sharp word reflected as an attack, every disagreement as a threat. Reactivity was the only option because they were seeing through a distorted lens.

But now, with a clearer mirror, they could distinguish between what was real and what was just a projection. The criticism wasn't a personal wound; it was a reflection of someone else's stress. The situation didn't require defense; it required presence.

What amazed them most was the freedom in that choice. They weren't trapped in automatic reactivity anymore. They didn't need to win the argument or prove their worth. They could choose a different response. And in choosing differently, they experienced something surprising: compassion.

Instead of walking away from the conversation angry and drained, they left with understanding. They saw their colleague not as an adversary but as another human carrying stress. And in that moment, the relationship grew stronger rather than weaker.

It wasn't about being perfect or never feeling defensive. It was about seeing clearly enough to pause, breathe, and reflect before reacting. That clarity opened a door to freedom. Freedom to choose, freedom to stay grounded, and freedom to respond with compassion. And in the end, the argument that once would have consumed an entire afternoon of frustration wasn't worth having at all. The moment of pausing in an argument, choosing not to snap back, is more than self-control. It's a glimpse of what life looks like when you see clearly. A clear mirror reflects reality without distortion: your worth isn't tied to criticism, and other people's reactions say more about their state than about you. When the mirror is clear, you can live with compassion and freedom instead of reactivity.

The Nature of Clear Seeing

To live with inner clarity means to see reality as it is, not as fear or judgment paints it. You no longer mistake temporary emotions for permanent truths.

You don't confuse someone else's stress with your identity. You notice what is happening in the moment without piling stories on top of it.

The difference between a distorted and an accurate reflection is the difference between looking through a dirty window and a clean one. Distorted reflection magnifies small issues into big problems, paints neutral events as threats, and convinces you that you are less capable than you are. Accurate reflection, by contrast, shows things as they are. Challenges as manageable, mistakes as lessons, and yourself as inherently worthy.

Clarity brings freedom from reactive living. When the mirror is clouded, you're at the mercy of every passing comment, mood, or setback. You react without choice. But when the mirror is clear, you gain the space to pause, reflect, and choose. That space is freedom, the freedom to respond with presence instead of being driven by automatic reactions.

Compassion as a Byproduct of Clarity

One of the surprising gifts of clarity is compassion. When you see clearly, you recognize that others are reflections too; their sharp words, cold tone, or impatient behavior are not definitions of who they are, but expressions of their current state. You stop seeing enemies and start seeing people who are struggling, just like you. Clear perception naturally gives rise to empathy. Instead of getting hooked by someone's anger, you can notice the fear or pain beneath it. Instead of resenting someone's criticism, you can recognize their stress or insecurity. This doesn't mean excusing harmful behavior; it means not letting it define you and not letting it block compassion.

Compassion heals relationships. When you meet others with understanding instead of defensiveness, the dynamic changes. Arguments dissolve more quickly, trust deepens, and connection grows. In families, workplaces, and communities, compassion clears the fog of conflict and creates space for collaboration and love.

Freedom in Reflection

Clarity also brings peace because it helps you detach from illusions. Many of the things that once consumed your energy, perfection, comparison, and approval, lose their grip when you see them for what they are: stories, not truths.

With clarity, you shift from reacting automatically to choosing responses. Instead of lashing out when criticized, you pause. Instead of collapsing under stress, you breathe. Each choice reinforces the truth that you are not a prisoner of reflexes.

This is the essence of living as the observer. You notice your thoughts, emotions, and surroundings without being ruled by them. You don't suppress them; you simply see them as passing

reflections. This observer perspective
grounds you in freedom. You are not your
passing thoughts. You are the one seeing
them.

A Reflective Life in the Modern World

Clear seeing is not just for meditation
cushions or quiet retreats; it's a practice
for daily life. At work, clarity allows you
to make decisions without being clouded
by panic or ego. In family life, it helps
you respond to conflict with patience
instead of anger. In communities, it opens
the door to dialogue instead of division.
Clarity also supports resilience in times of
crisis. When life throws unexpected
challenges (illness, loss, or major change),
a clear mirror helps you see the situation
as it is, not as your fears exaggerate it.
This doesn't erase pain, but it allows you
to move through it with steadiness and
perspective.

Reflection is not a one-time achievement.
It is a lifelong practice. Just as dust
gathers on a mirror day after day, life's

stresses and fears will continue to cloud perception. The work is not to keep the mirror permanently spotless, but to keep polishing it repeatedly. Each act of reflection brings you closer to clarity.

Reflection Tools

Pause Before Reacting

When you feel triggered, take one breath before responding. This tiny pause interrupts reactivity and gives space for choice.

Reframe Other People's Actions

When someone criticizes you, remind yourself: This is a reflection of their state, not my worth.

Compassion CheckIn

Ask: What might this person be feeling underneath their behavior? This softens judgment and opens empathy.

Daily Reflection Ritual

Set aside five minutes at the end of the day to ask: What did I see clearly today? Where was the mirror clouded? This builds awareness over time.

Key Takeaway

To see yourself clearly is to live with clarity, compassion, and freedom. It doesn't mean life becomes perfect, but it means you stop being trapped by distortions. You see yourself as whole, others as human, and challenges as opportunities. With a clear mirror, you can move through the world not as a reactor, but as a reflective, grounded presence.

Conclusion – One Mirror, Many Reflections

The journey of reflection begins within, but it does not end there. Every pause, every act of awareness, every choice made with clarity rather than confusion ripples outward into families, communities, and the world. When we reflect, we not only polish our own mirror, but we also contribute to the great mirror of humanity.

The Personal and Collective Journey

Reflection is a daily practice for individuals. A single mindful breath, a page in a journal, or a pause before reacting, these small practices become the foundation of a reflective life. They remind us that we are not our passing thoughts or feelings, but the awareness that notices them.

Reflection as a cultural force across traditions. From Hawaiian hoʻoponopono to Greek philosophy, from Buddhist mindfulness to African Ubuntu, reflection has always been more than self-improvement. It has been a way of restoring balance, preserving wisdom, and guiding communities toward harmony.

How both paths converge into shared wisdom. Personal reflection makes us clearer and kinder; cultural reflection makes societies more just and compassionate. Together, they form a timeless truth: that clarity is both a private gift and a collective responsibility.

The Truth of Interconnectedness

Why fighting others is fighting against ourselves. When we attack, exclude, or dehumanize others, we are striking at another reflection of the same human spirit. Reflection reveals that separation is an illusion; harming another is, in fact, harming ourselves.

Exclusion and division as distorted reflections. Prejudice, greed, and hostility arise from mirrors clouded by fear. They distort reality, tricking us into seeing enemies where there are only fellow travelers.

Connection is the natural state revealed by clarity. When the mirror is polished, when fear, judgment, and illusion are set aside, what remains is recognition. We see ourselves in others, and others in ourselves. Connection is not something we must build; it is what has always been there, waiting to be remembered.

The Vision of a Reflective Society

A world where every person has a place. In a reflective civilization, no one is left outside the circle. Every voice matters, every life has dignity, and every person belongs to the greater whole.

Systems built on compassion, fairness, and reflection. Justice is guided not by revenge but by restoration. Education teaches clarity alongside knowledge. Leadership values wisdom as much as strategy. Technology supports awareness rather than distraction.

Humanity shines brighter when each mirror is polished. When individuals reflect, their clarity adds to the collective. When societies reflect, humanity itself becomes more luminous, less driven by fear, more guided by compassion, and better prepared to face challenges together.

The Call to Action

Small personal practices that ripple outward. Begin with what you can: a pause before speaking, gratitude at the end of the day, listening deeply to someone else's story. These moments are not small; they are the seeds of transformation.

Living as examples of clarity and compassion. Reflection is not just what we do alone; it is how we live in the world. By embodying patience, empathy, and awareness, we become living mirrors for others.

Joining together to build a reflective future. The path forward is not about perfect individuals, but about communities and societies willing to reflect together. Each of us has a role in creating a future where clarity guides progress, compassion shapes systems, and humanity learns to see itself truly.

The One Mirror Practice

A daily ritual for clarity, connection, and reflection

Pause (Still the Water)

Take one slow, conscious breath.

Notice your body, your thoughts, your feelings — without judgment.

Imagine the surface of a lake growing calm.

Reflect (See Clearly)

Ask yourself one question:

"What is passing through my mirror right now?"

A thought? A feeling? A worry? A hope?

Recognize it as a reflection, not your identity.

Reframe (Polish the Mirror)

If the reflection is distorted (fear, anger, doubt), gently reframe:

"This is not me. This is passing through me."

If the reflection is clear (gratitude, love, insight), let it shine and expand.

Connect (See the Shared Mirror)

Think of one other person — a friend, a stranger, or even humanity at large.

Silently offer: "May you be clear. May you be at peace."

This step reminds us that the mirror is not only personal but collective.

Act (Shine the Reflection)

Choose one small action aligned with clarity — a kind word, a patient pause, or a step toward your values.

This is how reflection ripples into the world.

Why It Works

Step 1 calms the nervous system.

Step 2 builds self-awareness.

Step 3 strengthens resilience.

Step 4 deepens compassion.

Step 5 grounds reflection into daily life

Done in less than 5 minutes, this practice unites ancient wisdom and modern science, personal healing and collective harmony.

Closing Thought

We began this journey by exploring the mirror of the mind, how our inner state colors what we see in ourselves and the world. We end here with a vision: one mirror, many reflections. Each of us is a fragment of the whole, and together we form the great mirror of humanity. When we polish our own reflection, we help the world shine brighter.

The call is simple: pause, reflect, and live with clarity. From that small beginning, a new civilization can be born.

Final Words

We are each a mirror, sometimes clouded, sometimes clear, always reflecting. When we polish our own surface with awareness, compassion, and courage, we not only see ourselves more truly, but we also help the whole world shine more brightly. Every thought, every choice, every act of kindness ripples outward like light on water, touching mirrors we may never see.

The journey of reflection is not about perfection. It is about remembering: you are not your passing storms, not the shadows others cast, not the illusions of fear. You are the clarity behind them all.

And when enough of us live from that clarity, one breath, one reflection, one act at a time, a new kind of world begins to appear. A world where wisdom matters as much as knowledge, where compassion guides power, where each life is honored as part of the great whole.

This is not a dream for tomorrow. It begins in this moment, with you.

Look into the mirror. Breathe. And shine.

www.ingramcontent.com/pod-product-compliance
Lightning Source LLC
LaVergne TN
LVHW011910080426
835508LV00007BA/315